IBM-PC
in the laboratory

IBM-PC
in the laboratory

B.G.THOMPSON & A.F.KUCKES

*School of Applied and Engineering Physics, Cornell University
and Vector Magnetics Inc., Ithaca, NY*

Published by the Press Syndicate of the University of Cambridge
The Pitt Building, Trumpington Street, Cambridge CB2 1RP
40 West 20th Street, New York, NY 10011-4211, USA
10 Stamford Road, Oakleigh, Melbourne 3166, Australia

First published 1989
First paperback edition 1992
Reprinted 1994

British Library Cataloguing in Publication Data

Thompson, B. G. (Bruce Gregory), *1950–*
IBM-PC in the laboratory.
1. Physics. Applications of microcomputer systems
I. Title   II. Kuckes, A. F.
530′.028′5416

Library of Congress Cataloguing in Publication Data

Thompson, B. G.
IBM-PC in the laboratory / B. G. Thompson & A. F. Kuckes.
    p.   cm.
Includes index.
ISBN 0-521-32199-9
1. Physical laboratories—Data processing.   2. Physical
measurements—Data processing.   3. IBM Personal Computer.
I. Kuckes, Arthur F.   II. Title.
QC52.T46   1989
004.165—dc20   89-7079 CIP

ISBN 0 521 32199 9 hardback
ISBN 0 521 42867 X paperback

Transferred to digital printing 2004

Achilles: Before we start, I just was wondering, Mr. Crab—what are all these pieces of equipment, which you have in here?

Crab: Well, mostly they are just odds and ends—bits and pieces of old broken phonographs. Only a few souvenirs (*nervously tapping the buttons*), a few souvenirs of—of the TC-battles in which I have distinguished myself. Those keyboards attached to television screens, however, are my new toys. I have fifteen of them around here. They are a new kind of computer, a very small, very flexible type of computer—quite an advance over the previous types available. Few others seem to be quite as enthusiastic about them as I am, but I have faith that they will catch on in time.

Achilles: Do they have a special name?

Crab: Yes; they are called "smart-stupids", since they are so flexible, and have the potential to be either smart or stupid, depending on how skillfully they are instructed.

Acknowledgements
We wish to thank the students who struggled with us through the first semester of this course and Chuck Otis and Yvon Guy for their constructive comments on the notes. We also thank Katie Thompson for help with the Index and Table of Contents. Andy Draudt drew the pictures of the apparatus.

Note for AT class CPU users: The address 800($320) conflicts with the hard disk I/O address so place the I/O board base address at 544($220). References to I/O board addresses should be changed throughout the bool For instance, in program ADCTest (p. 14) the ADCRegister address becomes 556 or 544+12.

Contents

Contents

Contents

Contents

1 Introduction

The microprocessor has become commonplace in our technological society. Everything from dish washers to astronomical telescopes have chips controlling their operation. While the development of applications for computers has been in constant flux since their introduction, the principles of computer operation and of their use in sensing and control have remained stable. Those are the primary subjects of this book. Once a basic understanding of the principles has been built, further detailed knowledge can be acquired later as the need arises.

This book is designed to be accompanied by extensive laboratory work. Over the years the engineering curriculum has focused more and more on the lecture/recitation format. This has led to an ever increasing emphasis on theoretical developments and a loss of contact with the physical basis of engineering and science. The laboratory provides a vital experience in linking theory with physical reality. It also provides the satisfaction of building something and making it work.

Not all computers are suitable for laboratory use. Large mainframe computers are fast and can handle large amounts of data but are awkward to connect to laboratory equipment. At the other end of the scale, microprocessors are included in many laboratory devices but are programmed to perform only a restricted set of duties. Mini and microcomputers have enough speed and memory for all but the most demanding applications but yet are small enough to be dedicated to individual projects and therefore are widely used in the laboratory.

With the technological strides of recent years, microcomputers (or personal computers) have prodigious capabilities. Since they are also used in business, many languages and programs are available; some of which are even useful in a laboratory. With a single microcomputer, an engineer or scientist can acquire data and control an experiment, analyze the data, display the data and analysis as graphs or tables, and write a report or journal article. Remember that it can't do the thinking!

Microcomputers come with many built-in features. Those included depend on the designers' decisions as to what will sell the most computers. Since the demands of the laboratory are so varied, no computer when taken out of the box can hope to fulfil them. Hence to be useful, a computer must be able to change its capabilities after manufacture. This is done in two ways.

One is to provide a method of communication (serial or parallel) between the computer and the laboratory devices (analog to digital converters, voltmeters, etc.) and rely on the devices to be intelligent enough to communicate. Generally this means that the devices need a microprocessor built in which is preprogrammed to communicate with a certain protocol. The other way is to have slots (connectors) in the back of the computer so that circuit boards can be inserted to perform the desired tasks such as analog to digital conversion of serial communication. The computer can then be configured exactly as needed for a particular application. Even the video display or microprocessor can be changed when desired. Further, 'slot machines' are generally the least expensive way to computerize the laboratory.

The IBM-PC (upon which this book is based) and the APPLE IIe (which is the subject of a companion volume) are 'slot machines'. The APPLE IIe (and its predecessors, the APPLE II, the APPLE II+ and its successor APPLE IIGS) has proved its usefulness innumerable times. Its simple yet versatile architecture makes the computer easy to use and understand but limits its analysis and data volume capabilities. The IBM-PC is a newer design which is faster and has more memory but is slightly more complicated internally. They are both quite suitable for laboratory use. Beware of the APPLE IIc which is a slotless version of the IIe; it will not be able to accept the circuit cards which are necessary for the exercises in this book. IBM-PC clones (design copies) can be used instead of the IBM-PC as long as they have a slot where the data acquisition board can be placed and can run the programs. IBM-XT and AT designs can also be used. See Appendix A.

A computer can be treated as a black box which responds in a predictable way to an input; however, that type of use requires a complete knowledge of the possible inputs and responses. An understanding of how it works inside allows the user to figure out how the computer will respond to an input, or even if it can respond. The capabilities and limitations become transparent. Throughout the book a gradual understanding of what goes on inside a computer will be developed.

Other devices which are used are various sensors, analog to digital converters, digital to analog converters, timers, digital input and output devices, optical encoders, stepping motors, and analog amplifiers. They provide the interface between the computer's digital world and the physical phenomena being studied.

Turbo Pascal (version 4.0 or greater) is used for programming throughout the book. It has the advantage of being a structured language making the logical structure of programs more prominent. In addition, many students will have learned Pascal (or another structural language) in introductory programming courses. Turbo Pascal has additional capabilities which are convenient for data acquisition and control. These are the direct memory access instructions (such as Port and Mem) and the Inline statement which allows machine language statements to be inserted into the middle of Pascal programs. Turbo Pascal also has an integral editor for writing programs and is inexpensive. Turbo Pascal version 3 can also be used with minor changes in the text. The DOS program DEBUG is used to write some of the assembly language programs. A graphing/plotting program is also necessary; see Appendix B for guidelines in choosing one.

The text refers to the 8088 microprocessor. The details inside 8086/80286 based machines are different, however, programming is exactly the same at the level used in the book.

All of this computer work is done in the context of doing physics experiments. These experiments cover subjects not usually emphasized in introductory courses but which have a wide applicability. They show that, with computer control, conceptually sophisticated experiments can be performed with simple apparatus. In particular, the physics of activation temperature, heat diffusion and motion in fluids are explored.

1.1 How to use this book

In this book much of the programming material will be presented by way of example. Programs will be given from which you will be expected to deduce the essence of what is going on and thereby proceed to write your own programs. After using instructions in programs, the more precise description of instructions given in the manuals is easier to comprehend.

This book is written in a tutorial manner in which the exercises and experiments are distributed throughout the text. It would be nice to read up to an exercise and then sit down at the computer to do it. However, the time in the laboratory is short so that this becomes impossible. Before going to the laboratory, read through the text you expect to cover and organize your thoughts about what you will be doing. Also, jot down flow charts and write out programs which you will enter in the computer at the laboratory. Even if they do not run the first time they can be easily changed once the program is in a file. The essence of learning is going through the struggle of getting things to function properly, whether it be in writing programs, building experimental apparatus, or understanding theoretical descriptions.

Details of machine operation can be found in the operator's manual for the machine you are using. This is not a beginner's programming book; an introductory programming course or some experience in computer programming should precede the use of this book.

The appendices contain reference material and extended discussions. They are separated from the main text to improve the flow but contain important information and so should be perused once in a while.

1.2 Chapter summary

Chapter 2 begins with an introduction to the operation of the IBM-PC computer and Turbo Pascal programming. It also contains exercises in using the graphing program you have available.

Chapter 3 introduces the first Input/Output (I/O) device, the Analog to Digital Converter (ADC). It is used to measure the temperature/resistance characteristics of a thermistor. Further Pascal programming is used to do a least squares fit to the data. The I/O capabilities of an 8255 Programmable Peripheral Interface (PPI) chip are used to control a HEXFET switch on a heater to make a temperature controller.

In Chapter 4, simple time delays are used to generate control signals for a stepping motor and the effect of truncation errors is explored. The internal TimeOfDay clock of the IBM is used to make an internal timer.

Chapter 5 concerns an experiment in thermal diffusion. A heater at one end of a copper rod is turned on for a set interval under program control. The flow of this heat pulse down the rod is then measured at two locations for about 30 s. A theoretical model is fitted to these data to determine the thermal conductivity and heat capacity of copper. An analog amplifier is used to boost the signal from the thermistor to the ADC.

Chapter 6 is an introduction to assembly language programming and the internal structure of the IBM-PC. Simple programs are written using Debug and Turbo Inline statements. The efficiency of the Turbo compiler is examined. A Digital to Analog Converter (DAC) is used to make an X–Y plotter.

In Chapter 7, an experiment is constructed which measures the viscosity of glycerine by measuring the speed of a falling sphere. The physics of turbulent as well as smooth fluid flow is discussed. LEDs and photocells are used as position sensors to measure the speed of the sphere.

Chapter 8 introduces the concept of interrupt processing. Programs using software and hardware interrupts are written. Serial data communication is described and instated between two computers.

Chapter 9 contains various topics which are important but do not have a direct bearing on the experiments done in the previous sections.

2 Instrumentation structures and using the IBM-PC

The purpose of an instrument is to make measurements of a particular parameter in a physical process. This requires at least a sensor which responds to the parameter and a display which lets the user record readings which are in some way proportional to the parameter being measured. A thermometer is an instrument which indicates the temperature by quantitatively showing the expansion of a liquid with a temperature increase. A more complete description of the measurement process is shown in Figure 2.1. The arrows show possible but not necessary routes for the flow of information. The computer is able to do many of the tasks which were formerly done by separate units of an instrument. This lets the designer reduce the number of components required to a bare minimum as the experiments in this book show. Many times all that is needed is a sensor to translate the process into an electrical signal.

Another way to think of the computer is as an interface between the experimenter and the experiment (or the user and the measurement). It is able to translate the unintelligible signals from the sensor into a form which is understandable using human senses. One of the best ways of communicating information is by picture. 'A picture is worth a thousand words.' (In fact, it takes roughly a thousand words of computer memory to display a video graphics screen.)

Fig. 2.1. Instrumentation structures
Process: e.g. temperature as function of time, position as function of temperature.
Sensor: e.g. temperature or position converted to voltage.
Signal conditioner: e.g. amplifier, filter.
* Conversion: analog to digital.
* Storage/playback: e.g. silicon memory, magnetic, papertape.
* Representation: e.g. numbers, pictures (1000 words of memory!).
* Modeling: mathematical fit to data.
* Control: e.g. change temperature or position.
The computer has a part in all items with a *.

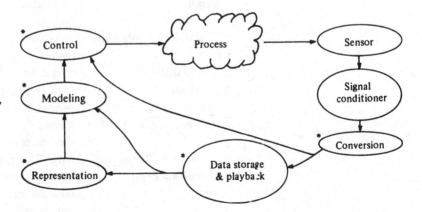

2.1 First program and graphs

Exercise 2.1.1   Starting out

(a) To get started: Insert the SYSTEM START disk into the drive unit marked 'A' and turn the computer power switch on. The disk is set up so that the computer will wake up in the Turbo Pascal system. To see what files are on the disk type 'F' (for File) and 'D' (for Directory) and then the RETURN key (CR, to accept the *.* wildcard mask). In addition to the Turbo files there are other files and programs such as 'COMMAND.COM', the operating system program (DOS) and a graphing program. Type the 'Esc' key twice or 'F10' once to get back to the main menu.

(b) Editing: To start writing a program, the program must be in the 'Edit' mode. Type 'E' at the main menu. The computer is now in Edit mode waiting for you to enter some text. 'F5' expands the Edit portion of the screen to full size. 'F5' again contracts it.

(c) First program: Type in program CALCCOS.PAS which follows. Each line should be ended with a CR. Use the Editing commands described in the Turbo manual to correct any mistakes you make. The comments can be abbreviated if desired but some should be included as reminders of why the statements are made.

```
program CalcCos;            {all words between brackets are comments}
                            {which are ignored by the compiler}

var                         {all variables used in the program must be}
                            {declared with a type, eg real, integer, string}
        y,x : real;         {all program statements end with a ; symbol}

begin                       {this announces that the main program
                               follows}
    writeln('    Radians    cosine');
                            {type header on the screen}
    x := 0.0;               {assigns the value 0.0 to the variable x}
    repeat                  {starts a repeat-until loop}
      y := cos(x);          {assign y, cos argument is in radians}
      writeln( x : 12 : 4, y : 12 : 4);
                            {type results on screen in format specified}
      x := x + 30*pi/180;   {step in 30 degree increments}
    until x > 2*pi;         {test end of repeat–until loop}
end.                        {end of program has a period not a
                               semicolon}
```

(*d*) Printing: One quick way to get a listing of the program is to use the Print Screen utility. Use 'F5' so that the full Edit screen is displayed. Make sure the printer is on and then press 'Shift PrtSC'. A copy of the screen should be printed on the printer.

(*e*) Running the program: To compile and run the program, either (i) exit the editor with 'F10' and press 'R' for Compile and Run or (ii) press 'ALT R' to go directly to Compile and Run. The computer will compile the program (that is translate the text in the Editor to machine instructions the microprocessor can understand). If there are no errors detected, the program will be run. An error in compiling will put you back in the editor near the place of the error so that you can correct it. After correction, run the program again.

The term compile means to translate the Pascal program to machine language. This is a necessary step because the program is written in text form (letters and digits) and the computer executes programs as binary code. The program can be stored as text and compiled each time it is used or in its binary form in which case it can be run from DOS, ie, outside the Turbo Pascal system. More on this in later chapters.

Also note that the computer makes a distinction between integer numbers (whole numbers) and real numbers (numbers which can have decimals). In general, integers are used for counting and reals for calculations. A different number of bytes is used to store integers and reals. A variable needs to be declared as one or the other so the computer knows how many bytes to access when the variable is used.

In order to use a program more than once, the computer can store the text on a disk and retrieve it later. The results of a program (such as the x,y list of program CalcCos) can also be stored and retrieved as a disk file. However, a disk must be initialized before it can be used. The system program disk is already initialized but it is better to have a separate disk to store your programs. Initialization is done with the DOS FORMAT.COM program. The following exercise shows how to use the program and how to use the disk to store programs and data text files.

Exercise 2.1.2 Saving programs and writing text files

(*a*) Formatting: First exit from Turbo by typing 'F', 'Q' at the main menu (or 'ALT F','Q' from anywhere else). If there is a program in the editor, Turbo will give a warning that the program has not been saved. Answer 'N', you don't want it saved. You should now be at the DOS 'A>' prompt. (If not type 'A :' CR.) Place a new disk in drive B and type FORMAT B: and then CR at the prompt. Be sure to use B: not A: since the FORMAT program erases all the data on a disk. You don't want to do that to the SYSTEM disk. The computer will spend a minute formatting the disk. When it is done take it out and label it with a felt tip marker.

(b) Another program: Put the new disk (call it the SAVE disk) back in drive B and type 'TURBO'. When Turbo has started, you need to tell it that disk operations will now be to drive B. Type 'F', 'C' for Change dir and then 'B : 'CR. Go to the editor and type in and run program WRITECOS . PAS which follows. This program will calculate the same data as program CalcCos but will also save it on disk. Take special note of the way the file statements work.

```
program WriteCos;         {calculates cosine and saves it on the disk}

var
    x,y : real;           {variables for the calculation}
    OutFile : text;       {variable for disk information assignment}
begin
    assign(OutFile, 'COS.DAT');
                          {make assignment to the disk information
                                variable}
    rewrite(OutFile);     {initialize the disk file for writing, this destroys
                           any file on the disk with the same name}
                          {use reset to open an existing file for reading}
    writeln('      Radians      Cosine');
                          {write to the screen}
    writeln( OutFile,'      Radians      Cosine');
                          {write the same header to the disk file}
    x := 0.0;             {same as before}
    repeat
        y := cos(x);
        writeln( x : 12 : 4, y : 12 : 4);
                          {write to the screen}
        writeln( OutFile, x : 12 : 4, y : 12 : 4);
                          {this writes the same information to the disk}
        x := x + 30*pi/180;
    until x>2*pi;
    close(OutFile);       {write the rest of the buffer to the file and}
                          {write an end of file marker}
end.
```

(c) Save the program: There are two ways to save a program file on the disk: if the file name which appears at the top of the edit window is the one you want, use 'F2' or 'ALT F','S' and the file will be saved under that name. Any previous file with that name will be renamed with a .BAK extension. If you want to rename the file while saving it, use 'ALT F','W' which will ask you for a new name to use. Use the second method here and name the file WRITECOS . PAS. Note if you just write WRITECOS the .PAS will be added automatically.

(d) Viewing text: To see the text created by the program you wrote, use the editor as follows. Type 'ALT F','L' for load and then the name of the file you wish to see, ie 'COS . DAT. Then enter the editor to see the numbers. To get the program back, use 'ALT F','L' again and enter the program name. Go back in the editor to make sure it is there.

(e) DOS commands: There is another way to view text. Exit the editor and type 'ALT L','Q' to exit Turbo. Now type 'TYPE COS . DAT' and the data file should appear on the screen. If it goes too fast, type 'Ctrl-S' to stop the text. Type any key to restart it. Another useful DOS command is PRINT. 'A : PRINT file' will print a copy of a file on the printer. Try printing your program and data file. Typing 'A : TURBO' will get you back to the Turbo Pascal program.

Graphing experimental data and mathematical expressions is an important aspect of the work outlined in this book. Many graphics programs are not suited to drawing graphs of data; they are written to draw pictures. On the IBM-PC there are several ways in which engineering graphs can be constructed. Turbo Pascal has some primitive graphing functions (PutPixel and LineTo) which allow a quick plot to be done from within a program. A library of functions can be constructed to add scales and text and make more elaborate plots. Another way to make graphs is to use a stand alone graphing program. The program reads data from text files created with an editor or a program and can plot it in several different ways with labeled axes, titles, etc. See Appendix B for a list of suitable programs.

Exercise 2.1.3 Simple graphing

Using the graphing method available on your computer, plot the curve $Y = X*X - 1$ from $X = -2$ to $+2$. You may need to write a program first to create a data file of the curve. Label the X and Y axis and make a dotted line grid at increments of 1 for X and Y. In addition, plot on your graph the data points X, Y as open circles for the following points:

X	Y
−1.8	3.5
−1.2	1.0
−0.5	−1.0
0.0	−1.3

0.5	0.3
1.2	0.5
2.0	3.5

The graphing program may have its own way of allowing you to enter this data or you could use the Turbo editor to make a file of the numbers.

Obtain a copy of the graph on the printer.

2.2 Addresses and data, RAM and ROM

Inside the IBM-PC there is an 8088 integrated circuit microprocessor which controls the operation of the computer. Connected to it are 16 primary address wires; eight of which also serve as data wires. There are also four secondary address wires. The 16 primary address wires allow the microprocessor to specify 65536 unique memory locations (addresses). During the data cycle of operation the eight data wires allow 256 unique numbers (or characters) to be represented. it is like a telephone system with 65536 telephone numbers and in which the caller can choose from 256 words to send a message. All the calls go through the central switchboard: the microprocessor.

The four secondary address wires allow addresses up to 16×65536 or 1 048 576. However, some special programming needs to be done in order to use those addresses. More about this later.

The address wires are connected to several different types of memory and to devices which allow communication between the computer and the outside world (for example the keyboard and screen). The microprocessor first places the binary representation of the location to be accessed on the address wires. Then after waiting for the computer circuits to select the unique location to which this refers, it either sends or receives a byte of data on the data wires during the data cycle. Along with performing some manipulation (eg, addition), this is all that a computer does.

Modern computers usually have several types of memory; early computers had only Random Access Memory (RAM). RAM is essential for any computer since the fundamental principles of computer operation require the Central Processing Unit (CPU) to repeatedly store and retrieve program instructions, data and memory addresses. The term 'Random Access Memory' means that it may be written to or read from in any order. A severe disadvantage of semiconductor RAM is that it doesn't remember anything after its power is turned off. Some computers have vital portions of their RAM protected by having a battery to provide the power in case of a power line failure.

Read Only Memory (ROM) has data stored in its memory cells at the time of manufacture which it retains permanently. It can be randomly accessed but that access is restricted to the read operation only. A ROM chip can be moved from one place to another without the data being lost as no power is needed to maintain data stored. There are several ROMs in the PC. One contains the Base System which is activated when the computer is turned on. Programs in the Base System initialize the computer and load the first programs from the disk. Another ROM contains the information on how to form letters on the video screen. Appendix C contains a description of how the PC memory is organized.

3 Thermistor experiments

In the first set of experiments you will make temperature measurements using a thermistor and an ADC. A thermistor is a device whose resistance varies with temperature. The ADC converts an analog voltage (continuous voltage levels) to a digital representation (discrete voltage levels) which can be read by the computer under program control.

3.1 Using the ADC

The ADC 0817 which is installed on the data acquisition card in the PC is an eight-bit converter; this means that the range of voltages from 0 to 5 V will be divided into $2^8 = 256$ parts. It is also able to select one of 16 input wires (channels) on which it will do the conversion. To use the ADC is quite easy: first an instruction is given to select the channel and initiate the voltage conversion; then a delay instruction is inserted to give the ADC time to work; finally the results of the conversion are read by a third instruction. Of the 16 analog input lines of the ADC, eight have been brought out by a cable to the prototyping board. Thus channels 0–7 on pins 1–8 of the Dual Inline Plug (DIP) connector on the prototype board. Appendix D has a more detailed description of how the ADC works.

The CPU of the IBM-PC (8088) uses a system of Input/Output (I/O) communication which is separate from the normal memory space addressing. There are 65536 locations which are dedicated to I/O programming. These are addressed by means of special instructions. In Turbo Pascal, the statement

Port[xxxx] : = yy;

will store the number yy in the I/O register xxxx; the statement

yy : = Port[xxxx];

will store the number read from the I/O register xxxx in the variable yy. The following exercise is an example of how these are used to communicate with the ADC.

Exercise 3.1.1 Using the ADC

To get the idea of how to use the ADC connect a 5 kΩ potentiometer across a 5 V power supply (Figure 3.1) and observe the voltage of the wiper (the center connection on the potentiometer) on the oscilloscope. NOTE: The ground lead of the oscilloscope probe should be connected to the ground of the system, ie, the green wire of the potentiometer. Never connect the oscilloscope probe ground to any point of a circuit which is not ground. Also, it is important that the ground of the power supply be connected to ground of the PC at all times.

Before connecting the wiper of the potentiometer to the PC observe the voltage of the wiper on the oscilloscope: set the scope trigger control to AUTO so that a continuous trace appears on the screen; be sure the small switch on the probe tip is set to 1x; set the vertical scale to 1.0 V/DIV and the VARIABLE knob to the CALibrated position; set a 0 V baseline by using the ground switch and vertical position knob on the scope.

Turn the knob of the test potentiometer back and forth, the voltage output of the wiper should vary between 0 and 5 V. Now

Fig. 3.1. Potentiometer connections and protoboard DIP plug.

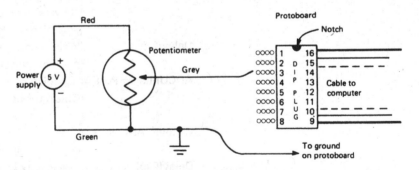

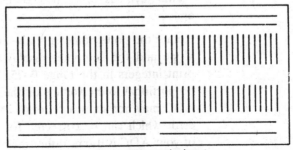

Protoboard connection layout

connect the wire from the wiper arm of the potentiometer to the channel 2 ADC analog input on the protoboard; this is pin 3 of the DIP connector.

Enter and run the following program:

```
program ADCTest;
{does continuous conversions of the voltage on channel 2}
{of the ADC on the John Bell Engineering Universal I/O board at}
{base address 800 ($320)}

uses crt;                      {this statement is necessary because}
                               {the crt unit contains the procedure}
                               {delay(n : integer); used below}
const
       ADCRegister = 812;      {the I/O register of the ADC}
       ChannelNo   = 2;        {the channel for conversion}
var
       ADCUnits : byte;   {conversion results in ADC units 0 to 255}
begin
  Repeat
  Port[ADCRegister] : = ChannelNo;
                       {Output channel no to be converted to}
                       {ADC register. This also starts the}
                       {conversion process}
                       {equivalent to Port[812] : = 2;}
       Delay(2);       {Wait at least 100 microsec for the ADC to}
                       {perform the conversion and have valid data}
  ADCUnits : = Port[ADCRegister];
                       {Obtain the coversion results from}
                       {the ADC and store in a variable}
       writeln(  ADCUnits : 8  );
                       {display results on screen}
       Delay(500);     {wait approx ½ s before doing the next}
                       {so the screen doesn't scroll too fast}
    until Keypressed;   {loop continuously until a key is pressed}
end.
```

Rotate the potentiometer shaft and note how the voltage on the scope and the computer display changes. The program should print integers in the range 0–255 on the video screen which are proportional to the voltage on the potentiometer. Your particular ADC may not show a count of 255 for 5 V. This is a calibration error which can be corrected for by determining that the range of your ADC is 0–*xxx* rather than 0–255 for a 0–5 V input.

The potentiometer is an example of a zero-order instrument; ie, it is a transducer whose output is in direct proportion to its input: $V_{out} = KV_{in}$ where K is the static sensitivity or calibration factor. A perfect zero-order instrument will produce at its output the exact replica of the input signal with only a scale or units change. Of course no instrument or transducer can live up to the perfect response represented by a mathematical formula; all instruments have a range of input values over which tolerable errors occur. It is the responsibility of the designer to determine this range and report the tolerance in the instrument specifications and the responsibility of the user to pay attention to them.

Exercise 3.1.2 Programming the ADC

(a) Modify the program and potentiometer connections so that the voltage on channel 5 is read and displayed. Determine what the ADC reading is for the maximum voltage, and what the maximum voltage is.

(b) Modify the program and add a second potentiometer so that the program reads the voltage on channel 0 and then, as soon as possible, the voltage on channel 5, the two results should be displayed on the crt on a single line with a few spaces between the two measurements. Make the program also compute the actual voltages and print them too. The program should make 25 measurements of this kind and then halt. When you get everything running, print out the results on the printer. Also make a printed listing of the program you have written and save the program on your disk.

3.2 ADCs

ADCs come in many sizes and flavors each with a range of usefulness. The following is a description of the most important considerations for choosing and using them.

An ADC has a defined range of input voltages (for example 0–5 V) which it can accurately convert. This range is divided into a number of equal sized pieces (voltages). The integer number output by the ADC corresponds to the number of these which equal the input voltage at the time of the conversion. Figure 3.2 shows how a three binary bit converter converts an input signal to a digital number. Using an ADC is like using a ruler which is graduated in say $\frac{1}{8}''$ markings. All measurements are then made to the nearest $\frac{1}{8}''$. Also it can only measure lengths which are less than the length of the ruler. (You can hop-frog a ruler but you can't do that with an ADC.)

The goal of digital measurements is to get an accurate representation of the input signal. In order to do this the ADC must be able to resolve voltage differences which are significant in the measurement being done. That is, the

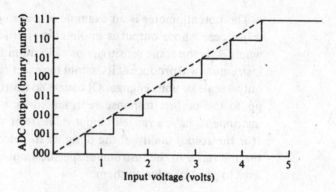

Fig. 3.2. Output of a three-bit ADC with input range of 0–5 V. Dashed line is the ideal, solid line the actual response.

input voltage range of the ADC must be divided into enough pieces by the digitizer that the voltage change represented by each piece is smaller than the accuracy needed. In the laboratory, the 8-bit converter breaks up the input analog voltage range into $2^8 = 256$ pieces so that the resolution of the converter is $1/256 = 0.4\%$ of the full range or $(5 - 0)/256 = 0.019$ V. Digital audio recording systems use 16-bit converters so that the digitization process is not audible on playback. The ear is a very sensitive detector.

Most of the ADCs on the market are 8, 10, 12, or 16-bit converters. Those above 12-bit require extra care in use since the digitization level is below 1 mV. Extraneous noise from computers or other circuits can creep into the desired signal. Common input voltage ranges are (0 to 5), (−5 to 5) and (−10 to 10). External electronics can be used to shift the voltage from a sensor into the proper range.

Exercise 3.2.1 ADC and sampling

(a) The resolution in voltage of an ADC is $\Delta V/2^n$ where ΔV is the total input range and n the number of bits of the digital output. What is the resolution of a 13-bit ADC with input range of +5 to −5 V? Express your answer in millivolts.

(b) Since the amplitude of an analog signal can be adjusted by an amplifier circuit to fill the input range of the ADC, the resolution can be better described by the dynamic range; this is the ratio of the maximum to the minimum voltage measurable by the ADC. The maximum is the ADC input range and the minimum is the resolution calculated above. What is the dynamic range of the 13-bit ADC? The 8-bit used in class? The ratio is usually expressed in decibels (dB), eg, $DR = 20 \log_{10}(\text{ratio})$ in dB. Give your answers in both forms, as a ratio and in dB.

Another method of converting an analog signal to digital is to input the signal to an electronic circuit (a Voltage Controlled Oscillator or VCO)

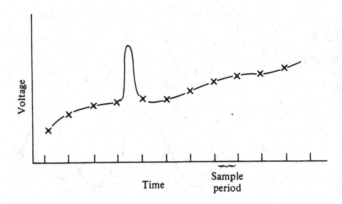

Fig. 3.3. Sampling the signal with an ADC at a rate which is too slow. The peak is missed.

whose output is a frequency which is proportional to the amplitude of the input voltage, $f_{out} = f_0 + KfV_{in}$. The computer can then measure the frequency of the signal by measuring the time for one cycle of the waveform. To work properly the rate at which the analog input voltage varies must be much less than the frequency output and so the VCO is used for slowly varying signals. The accuracy of this method is also limited.

In order to measure a signal accurately, the rate at which the measurements are taken (the sampling rate) must also be considered. This must be fast enough that all the frequencies contained in the signal can be reproduced. As a quick illustration of the problem, the signal peak in Figure 3.3 will be totally missed if the sampling is done at the time marked with crosses.

As Fourier (1768–1830) showed, any signal can be considered as a superposition of sinusoidal signals of various frequencies. These frequencies generally range from zero (DC) to some maximum frequency, f_{max}, which depends on the physical characteristics of the system generating the signal. The fundamental frequencies of piano range from 27 to 4200 Hz. But the overtones (harmonics) go to much higher frequencies.

In order to reconstruct the original signal from a sampled signal accurately, the ADC sample rate should be at least twice the highest frequency in the input signal, f_{max}. This result is known as the Sampling Theorem and was formulated by Shannon in 1949 building on earlier work by Nyquist (1924). Note that it reads 'the highest frequency in the input signal' not 'the highest frequency of interest'. Even if you are not interested in higher frequencies in the input signal, they must be sampled correctly. If they are sampled at a rate which is less than $2f_{max}$ (the Nyquist frequency), they will masquerade as lower frequency signals (Figure 3.4). This is called aliasing. A good rule of thumb is use a sample rate of at least $2.5f_{max}$. Electronic filters (like the treble and bass controls on a stereo) can be used to limit the frequency range of signals so that the sampling rate can be lowered.

As an example, in the digital recording of music, the audio frequency range of 20–20 000 Hz must be faithfully sampled. Since $f_{max} = 20\,000$ Hz,

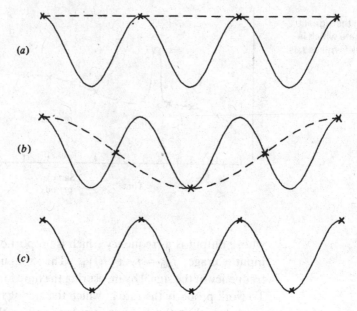

Fig. 3.4. The solid curve is the input waveform. The dashed curve is the waveform reconstructed from the sampled data; (*a*) 1 sample per cycle, (*b*) 1.5 samples per cycle, (*c*) 2 samples per cycle (the Nyquist frequency).

the Nyquist frequency is 40 000 samples/second. The actual rate used is 48 000 Hz. An ADC which converts the signal in less than 20 μs is needed.

ADCs come in a wide variety of speeds, from the low-power devices which convert in milliseconds to fast (less than 10 μs) ones. The first are used in battery operated equipment such as digital multimeters. The fast ones are used in audio and video digital systems.

Exercise 3.2.2 Audio digital sampling

An eight channel digital recording studio wants to faithfully record the audio spectrum from 20 to 20 000 Hz. What must be the sample rate for each channel? The studio wants to use a single multiplexed 16-bit ADC to digitize the signals (the ear is a sensitive detector); what is the maximum conversion time the ADC can have? It is found that ADCs this fast are only available to the military (and at military prices!), but there are some available which are three times slower; how can the system be changed to accommodate slower ADCs and still have the full frequency capability?

There are various output formats for the data coming from ADCs to the computer. This is usually not a large concern when buying them since the computer can convert any format into the one most suitable for its use. Table 3.1 shows some standard output codes for an eight-bit converter with an

Table 3.1 *Comparison of ADC numbering systems*

Input volts	2s complement	Offset binary	Sign bit
+10	0111 1111	1111 1111	1111 1111
+10 − LSB	0111 1110	1111 1110	1111 1110
. . .			
+1 LSB	0000 0001	1000 0001	1000 0001
0	0000 0000	1000 0000	1000 0000
−1 LSB	1111 1111	0111 1111	0000 0001
. . .			
−10 + LSB	1000 0001	0000 0001	0111 1110
−10	1000 0000	0000 0000	0111 1111

input range of −10 to 10 V. One LSB (Least Significant Bit) represents $20/256 = 0.078$ V.

3.3 Thermistor resistance vs. temperature characteristics

The first real application to which the ADC will be put is to measure the resistance variation of a thermistor with temperature. The electrical resistance R of conductors (metals) increases with increasing temperature. This is a result of the change in the mean free rate between collisions of the free electrons in the conductor with the lattice (stationary ions). As the system heats the increased amplitude of thermally generated lattice vibrations (phonons) results in an increased R. Yet for a thermistor, R *decreases* with increasing temperature according to the relation $R = R_0 \exp(T_0/T)$. (You may want to make a quick plot of $y = \exp(1/x)$ to see the rough behaviour of this function.) In this expression, R is resistance (ohms); R_0 is a resistance value corresponding to infinite temperature; T_0 is an activation temperature (K); T is the absolute temperature (K) (0 °C = 273.16 K). The difference is that the thermistor is made from semiconducting materials. In a conductor, every atom donates one or more electrons to the conduction electrons and thus the number of conduction electrons is fixed at a rather large number $\approx 10^{22}/\text{cm}^3$. In a semiconductor the electrons are more tightly bound to the atoms. The energy required to liberate electrons from the atoms is $Eg = k_B T_0$ where k_B = Boltzmann's constant. The probability of an electron being liberated from any given atom by thermal agitation is $p = \exp(-E_g/k_B T) = \exp(-T_0/T)$. Thus the number density of free electrons in a semiconductor varies as $n = n_0 \exp(-T_0/T)$. Note that for $T \ll T_0$, n goes to 0 and the semiconductor becomes an insulator. Since the resistance of a conductor depends inversely on both the number of charge carriers and the mean free path of the carriers, the rapid variation of n with T dominates the resistance of a semiconductor overriding the temperature effect on the mean free path, which can be ignored to a good approximation.

Exercise 3.3.1 Thermistor mathematical models

To show that this last statement is true consider two models of thermistor behavior

$$R_1 = R_0 \exp(T_0/T)$$

and

$$R_2 = AT \exp(T_0/T)$$

where R_0, T_0 and A are constants and R_2 includes the effects of the mean-free-path variation with temperature. Plot $\log(R_1)$ vs. $1/T$ for $T_0 = 3000$ K and $R_0 = 0.02\ \Omega$ in the temperature range 280–400 K. Now plot $\log(R_2)$ for the same T_0 and adjust A so that $R_2 = R_1$ at 300 K. Show mathematically that R_1 *should* and R_2 *should not* plot as a straight line on this type of plot. Despite this R_2 does appear to be a straight line in this temperature range and so it can be modeled with the equation $R_2' = R_0' \exp(T_0'/T)$. From the graph, find T_0'. Where does the behavior of R_2 start to differ significantly from R_1, at low temperatures or high?

The thermometer/thermistor protoboard has the circuit diagrammed in Figure 3.5. NOTE: when any wiring is done or changed be sure to turn off the power supply. Even though the wiring is very simple it is still a good idea to become accustomed to using wire color codes to help you. Red is used for positive power supply connections, green for ground and the standard electronic color code (Table 3.2) if there is an easy correspondence to data line numbers. Taking the time to do this will make it easier to trace the circuit to find errors when something doesn't function correctly. In addition it is much easier to find test points when a scope or other test instruments are to be used. The voltage across the thermistor should be read into channel 0 of the ADC (ADC0). The push button switch makes it possible to turn the heater on and off manually.

Fig. 3.5. Thermistor and heater circuit.

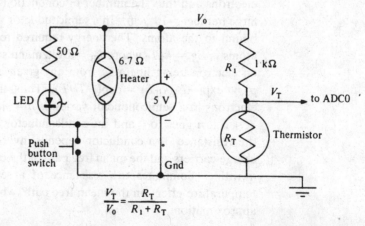

$$\frac{V_T}{V_0} = \frac{R_T}{R_1 + R_T}$$

Table 3.2 *The standard electronic color code and resistor identification*

	Color code
0	Black
1	Brown
2	Red
3	Orange
4	Yellow
5	Green
6	Blue
7	Violet
8	Grey
9	White
5%	Gold ⎫
	⎬ Tolerance
10%	Silver ⎭

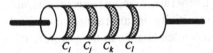

$$C_i \quad C_j \quad C_k \quad C_l$$

$$R = C_i C_j \times 10^{C_k} \pm C_l \%$$

Exercise 3.3.2 Specific heat and power

This exercise is a warm up for Chapter 4 and uses the thermistor circuit but not the computer. The specific heat of a substance is the ratio of the amount of heat added (ΔQ) to the corresponding temperature rise (ΔT) per unit mass (m):

$$C = \Delta Q / m \Delta T \tag{3.3.1}$$

The power P is defined to be the change in heat with a change in time,

$$P = dQ/dt \tag{3.3.2}$$

so the amount of heat added to the aluminum block by the heater is the power times the time: $\Delta Q = P\Delta t$. Where the power is the voltage drop v times the current i, $P = iv$ or by using Ohm's law $P = v^2/R$ for a resistance R.

(a) While pushing the button, determine how rapidly the temperature rises (degrees/second). By estimating the mass of the aluminum block, *estimate* its specific heat. In the *CRC Handbook of Chemistry and Physics* it lists the specific heat of aluminum as 0.215 cal/g °C and the specific gravity as 2.702 g/cm³. Convert the specific heat to

SI (kg-m-s) units and compare with your rough results. Where does error enter this estimate?

(b) Also measure the rate at which it cools and calculate the heat lost per unit time (the power output) due to conduction and convection. Is this result significant for the measurement made in part (a)?

(c) When you release the button, why doesn't the temperature stop rising immediately?

The thermometer used to measure the temperature of the block has thus far been considered a zero-order transducer (like the potentiometer in Section 3.1). In reality it takes a finite amount of time for the mercury in the glass bulb to heat up in response to the increase in temperature of the block. It is thus a first-order instrument whose response is determined by the differential equation: $\tau(dT_{out}/dt) + T_{out} = KT_{in}$ where T_{out} is the change in the reading on the thermometer (output), T_{in} is the change in the block temperature (input), τ is the characteristic response time, and K is the static sensitivity or calibration factor. The solution of this equation for the case of a sudden change in temperature of the block is $T_{out} = KT_{in}[1 - \exp(-t/\tau)]$ whose graph is shown in Figure 3.6. This shows that if changes happen quickly enough, the thermometer response does not keep up and the readings will be in error. Notice that if $t = \tau$, the temperature has risen to $[1 - \exp(-1)]$ or about $\frac{2}{3}$ of the step input change (see Figure 3.6). This provides a quick way to estimate τ.

Higher-order instrument response characteristics are also common in instrumentation systems. For example, a damped spring used for weighing objects or as an accelerometer is usually modeled by a second-order

Fig. 3.6. Graph of the time response (dashed line) of a thermometer (first-order transducer) to a step temperature change (solid line) in the surroundings.

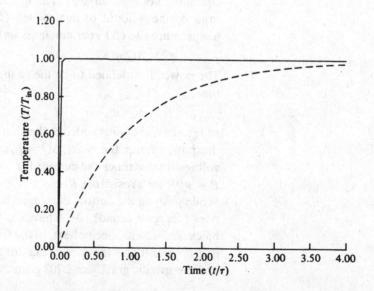

differential equation. Three parameters are then needed to predict the response to a particular input: the calibration K, the damping constant δ, and the resonant frequency f_0. Each frequency of the input signal is affected in a different way as it passes through the system. In systems of order two or higher a graph of the gain or calibration factor as a function of frequency is useful for designing instruments using a particular transducer. The frequency response characteristics of a cassette tape or a stereo phonograph are often displayed in their advertising literature. The frequency and phase vs. frequency for a second-order transducer is shown in Figures 3.7(a) and (b). Note that at the resonant frequency a large response can occur if the damping is weak.

In the laboratory, both the thermometer and the thermistor are first-order transducers and so have finite time responses to a change in temperature.

Fig. 3.7. Frequency response of a second order transducer for the damping constant $\delta = 0.1$ (solid line) and for the damping constant $\delta = 1.0$ (dashed line). (a) Amplitude vs. frequency: (b) phase vs. frequency.

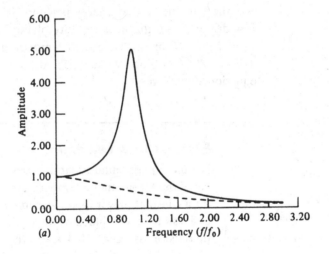

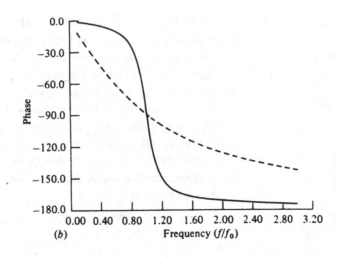

However, the time constant of the thermometer is much larger and so it dominates the response of the system. The time constant of the thermometer can be estimated to be about 1.5 s by watching the temperature reach equilibrium after the power input is stopped (the button is released). In the following, the purpose is to estimate the temperature lag of the thermometer behind the block temperature for a constant power input (ie, when the button is kept pushed down, how far behind the actual temperature is the measured temperature?)

As before, the differential equation of the response of a first-order temperature transducer is

$$\tau \, dT_o/dt + T_o = T_i$$

where T_o is the temperature measured minus the initial temperature of the system and T_i is the temperature change input to the system. The power input to the block is $P = \Delta V^2/R$ with R the heater resistance. The power is also the amount of heat energy per unit time which goes into the block $P = dQ/dt$. Since the heat capacity of the block is $C_V = dQ/V \, dT$, then $C_V = P \, dt/V \, dT$ or the change in temperature of the block with time is $dT/dt = P/C_V V$. Since $dT/dt = dT_o/dt$, substituting into the differential equation above gives

$$T_o - T_i = \tau \, P/C_V V$$

Exercise 3.3.3 Lag time

For the experimental apparatus used in the laboratory, estimate the lag of the thermometer temperature behind the block temperature for a constant power input. Assume $\tau = 1.5$ s for the thermometer and estimate the block volume. Do the same estimate for the thermistor using $\tau = 0.4$ s for the time constant. What is the difference between the thermometer lag and the thermistor lag?

Exercise 3.3.4 Thermistor resistance measurement

Write a program which will allow you to enter manually a tempera-ture reading you observe on the thermometer using a Readln; statement and which will then read the thermistor voltage by using the ADC. Check the voltage readings printed out on the crt screen with those which you get with the oscilloscope. Make a printout of the program when it works. Once you get this working write a few additional statements so that the actual resistance of the thermistor is computed and printed. Follow the flow diagram in Figure 3.8. The resistance can be calculated from the voltage divider relationship shown on Figure 3.5:

$$\frac{V_T}{V_0} = \frac{R_T}{R_1 + R_T} \tag{3.3.3}$$

Fig. 3.8. Flow chart for Exercise 3.3.4.

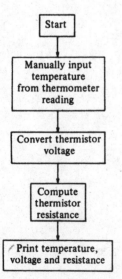

Make a quick check of the computer code by doing a calculation by hand.

Exercise 3.3.5 Data arrays

Modify the program in Exercise 3.3.4 so that the computer makes a series of measurements and stores them in arrays, ie, Temp[I], Res[I]. These symbols mean that measurement number I had a thermometer reading of Temp[I] and a resistance measurement of Res[I]. Print the whole array after the last measurement is made. To get out of the input loop use some absurd value of the temperature (say 0 or 1000) as a flag that no more input is desired. (Make sure that this last value is not included in the data.) Shortly, you will add additional steps so that these data can be stored on the disk as a data file.

3.4 Making and retrieving sequential data files

Saving and retrieving data from a disk is a valuable part of what a computer can do. In program WriteCos the values of the calculation were written to the disk as a text file so that they could be used by a graphing program. Here the temperature and resistance data will be written by one program and read by another. The parts of the programs which write or read the data are written as procedures to separate them logically from the rest of the program. To supplement the comments please read the reference manual sections on procedures and data files. First the procedure for writing the arrays.

Exercise 3.4.1 WriteArrays procedure

Write the following procedure and save it on disk as WRARRAY.PAS.

```
procedure WriteArrays(  NData : integer;
                        X,Y : DataType;
                        OutFilename : FilenameType );
{to write the data arrays X and Y to the disk file named}
{OutFileName}
{NOTE: the following statements must be in the calling program}
{type                                              }
{     DataType = Array[1..100] of real;            }
{     FilenameType = String[40];              }
{                                              }
{so that the arrays and name can be passed to the procedure.}
```

```
    var
         OutFile : text;      {for assignment to a disk file}
         i        : integer;  {for loop}

begin
   assign(OutFile, OutFilename);    {assign disk file}
   rewrite(OutFile);                {prepare disk for writing}
   for i : = 1 to NData do {write pairs of data}
     writeln(  OutFile, X[i] : 12 : 4, Y[i] : 12 : 4  );
   close(OutFile);                  {write buffer and close file}
  end;                              {end of procedure WriteArrays}
```

Note that preparing the disk file to accept data is a two-step process: Assign and Rewrite. Assign prepares a buffer space in memory and a file variable for reference. Rewrite prepares the disk drive for the data to come. Any existing file with the same name will be overwritten when Rewrite is executed. The Close statement is important because the data is not always written to the disk immediately. It is stored in a memory buffer until the buffer is full and then sent to the disk. Frequently at the end of a program the last data bytes are still in the buffer. The Close statement will ensure that the bytes are written to the disk and that an EndOfFile marker is placed on the disk.

The procedure WriteArrays cannot be compiled and executed as it stands; it needs to be in a program.

Exercise 3.4.2   Program TestWrite

Write and run the following program to test the WriteArrays procedure. Look at the data file 'XSQ.DAT' with the editor to see that it is correct.

```
program testwrarray;
{to test the WriteArrays procedure}
type
      DataType = Array[1..100] of real;
                           {these are for passing variables}
      FilenameType = String[40];
                           {to the procedure WriteArrays}
const
      N = 10;               {number of array elements to calculate}
var
      i : integer;         {a counter}
      x,xsq : DataType;    {arrays to calculate and write}
```

```
{bring file WRARRAY.PAS into program by placing cursor at
the following blank line and using the command Ctrl-K R}
begin
  for i := 1 to N do begin
    x[i] := i;                  {calculate arrays}
    xsq[i] := i*i;
    writeln( x[i] : 12 : 4, xsq[i] : 12 : 4 );
                                {write arrays to screen}
  end;
  WriteArrays( N, x, xsq, 'xsq.dat' );
                                {write arrays to disk file}
end.
```

The critical (and useful!) part of making files is being able to read the data back into a program. The following procedure will do this. Notice that the number of data pairs is obtained from the file by using the EOF procedure to test for the end of a file. Also, Reset is used to prepare the disk file for reading instead of Rewrite which would erase the existing file.

Exercise 3.4.3 ReadArrays procedure

Write the following procedure and save it on the disk as REARRAY.PAS.

```
procedure ReadArrays( var NElements : integer;
                      var X,Y        : DataType;
                      InFilename     : FilenameType);
{to read the data arrays X and Y from the disk file named}
{InFilename}
{NOTE: the following statements must be in the calling program}
{ type                                                        }
{      DataType = Array[1..100] of real;                      }
{      FilenameType = String[40];                             }
{                                                             }
{so that the arrays and name can be passed to the procedure.}


var
    InFile : text;        {for assignment to a disk file}
    i      : integer;     {counter for loop}


begin
  assign(InFile, InFilename);
                          {assign disk file}
  reset(InFile);          {prepare disk for reading}
                          {note this is different than rewrite!}
```

```
    i := 0;                    {initialize counter}
    while not EOF(InFile) do begin
                               {start loop but test for EndOfFile}
      i := i + 1;              {bump counter}
      readln( InFile, X[i], Y[i] );    {get data into arrays}
    end;
    NElements := i;            {has proper count when it exists while loop}
    close(InFile);             {close file}
  end;                         {end of procedure ReadArrays}
```

Exercise 3.4.4   Program TestRead

Write and run the following program to test the ReadArrays procedure.

```
program testread;
{to test the ReadArrays procedure}

type
DataType = Array[1..100] of real;
                           {these are for passing variables}
FilenameType = String[40];
                           {to the procedure WriteArrays}

var
    i : integer;           {a counter}
    NData : integer;    { the number of array elements obtained}
    x,xsq : DataType;   {arrays to read from file}
{another way to use an existing procedure file is to direct the}
{compiler to insert it at the appropriate place}
{the following statement with the braces included}
{will do this provided it is on a line by itself}

{$I  REARRAY.PAS}

begin
  ReadArrays( NData, x, xsq, 'XSQ.DAT' );
                           {read data arrays}
  for i := 1 to NData do
    writeln(  x[i] : 12 : 4, xsq[i] : 12 : 4 );
                           {write arrays to screen}
end.
```

Exercise 3.4.5 Temperature and thermistor resistance data file

Modify your thermistor program (Exercise 3.3.5) so that temperature and thermistor resistance arrays are recorded on a disk. Do enough tests to be confident that the program you have written generates a disk file and that you are able to read it back.

Using the manual on/off switch on the heater, make a series of measurements of temperature and resistance (about 10–15) of the heater block between room temperature and about 100 °C. Record them as a disk file. Read the data back and print them so that you know you have them.

3.5 Plotting the experimental data

Exercise 3.5.1 Thermistor data plot

Using the graphing method available to you, plot the thermistor resistance vs. temperature. It is necessary to use degrees Kelvin in later exercises so scale the data appropriately and plot the temperature in degrees Kelvin.

The value of graphical plots is that they are capable of displaying and conveying much information very quickly. One obvious weakness of the linearly scaled display of the resistance vs. temperature plot which you have made is that it is difficult to get a good display of the lower values of resistance. When the numerical value of a parameter to be plotted spans a large range, scaling the axis logarithmically is very useful. On a linearly scaled axis each increment of length is proportional to an increment of the parameter being plotted. On a logarithmically scaled axis each increment of length is proportional to the fractional change in the value of the parameter. (If $y = \log(R)$, then $dy = dR/R$.) Often it is more significant to note the fractional change in a parameter than the change in the value of the parameter itself. For example, when plotting stock exchange prices and their change in time, it is much more useful to plot the stock prices on a logarithmic ordinate scale than a linear one.

Exercise 3.5.2 Logarithmic plot

Modify your plot to use a logarithmic scale on the ordinate. Note that using a logarithmic scale is different from plotting logarithmic values on a linear scale although the shape of the resulting curve is the same. Some plotting programs do not have the capability of displaying logarithmic scales so logarithmic values on a linear scale will have to do.

This logarithmic plot is a very useful one to display the resistance of a thermistor vs. temperature for purposes of manually determining the resistance for a given temperature. However, for comparison with mathematical theory it is better to use a different plot. The form of the plot is determined by the particular phenomena being studied. As shown in Section 3.3, the variation of the resistance of a thermistor can be written as:

$$R = R_0 \exp(T_0/T) \tag{3.5.1}$$

where R_0 (Ω) and T_0 (K) are constants. To display graphically the extent to which the measured dependence conforms to this theoretical dependence, it is useful to plot the resistance vs. temperature using a scale such that the resulting plot becomes a straight line. This is easily done by taking the natural logarithm of the resistance for the 'linear' ordinate length and $1/T$ for the 'linear' abscissa length scale. Taking the logarithm of Equation (3.5.1) gives

$$\ln(R) = \ln(R_0) + (T_0/T) \tag{3.5.2}$$

and by setting

$$y = \ln(R) \quad A = T_0 \quad x = 1/T \quad B = \ln(R_0) \tag{3.5.3}$$

Equation (3.5.2) becomes

$$y = Ax + B \tag{3.5.4}$$

which is a straight line. (You'll notice, on close inspection, that the previous plot in Exercise 3.5.2 is not a straight line.)

Exercise 3.5.3 Linearized thermistor data plot

Make a modified data file which has resistance and 1/temperature data pairs and plot the resistance vs. 1/temperature using a logarithmic scale for the resistance. Be sure that the temperature is in units of Kelvin^{-1}. Check to see if your data conforms to the model, Equation (3.5.1).

3.6 A least squares fit to the data

Finding good values for the parameters R_0 and T_0 in Equation (3.5.1) is important for investigating the physics of the device; having good values for them is also important for making the temperature controller which you will be shortly called upon to do. By finding values for A and B from the linear plot of Exercise 3.5.3, values for R_0 and T_0 can be easily calculated via Equation (3.5.3). This can be done graphically or by a least squares fit of the data.

In doing the experiment, you have acquired data at a sequence of values of temperature T_i or alternatively $X_i = 1/T_i$. Each of these temperatures T_i yielded an experimental resistance value R_i or alternatively $Y_i^{ex} = \ln(R_i)$. Equation (3.5.1) yields a theoretical resistance value R_i^{th} for each temperature, ie, for each X_i a theoretical value $Y_i^{th} = \ln(R_i^{th})$ is given. The task is to find values for A and for B to minimize the error between the experimental and theoretical values, $E_i = Y_i^{th} - Y_i^{ex}$. A common type of analysis minimizes the sum of the squares of the individual errors. Calling the total square of the error E_T, we get

$$\left. \begin{aligned} E_T &= \sum_i E_i^2 \\ E_T &= \sum_i (Y_i^{th} - Y_i^{ex})^2 = \sum_i (AX_i + B - Y_i^{ex})^2 \end{aligned} \right\} \quad (3.6.1)$$

To minimize this error with respect to the parameters A and B we take derivatives with respect to A and B and set them to zero:

$$\left. \begin{aligned} \partial E_T/\partial A &= 0 = \sum_i 2X_i(AX_i + B - Y_i^{ex}) \\ \partial E_T/\partial B &= 0 = \sum_i 2(AX_i + B - Y_i^{ex}) \end{aligned} \right\} \quad (3.6.2)$$

Taking A and B out of the summations and collecting terms gives

$$\left. \begin{aligned} AS_{XX} + BS_X &= S_{XY} \\ AS_X + BS \ \ &= S_Y \end{aligned} \right\} \quad (3.6.3)$$

where

$$\left. \begin{aligned} S_{XX} &= \sum_i X_i^2 \qquad S_Y = \sum_i Y_i^{ex} \\ S_X &= \sum_i X_i \qquad S = \sum_i 1 \\ S_{XY} &= \sum_i X_i Y_i^{ex} \end{aligned} \right\} \quad (3.6.4)$$

Then solving for A and B

$$\left. \begin{aligned} D &= SS_{XX} - S_X^2 \\ A &= \frac{SS_{XY} - S_X S_Y}{D} \\ B &= \frac{S_{XX}S_Y - S_{XY}S_X}{D} \end{aligned} \right\} \quad (3.6.5)$$

Exercise 3.6.1 Least squares fit to data

(a) Write a procedure to find values for A and B using a linear least squares fit to arrays of data X[i] and Y[i].

(b) Use the procedure in part (a) to obtain the model fit to your resistance and temperature data and obtain values for T_0 and R_0.

(*c*) Plot the theoretical fit as a line together with your experimental
values as open circles using a logarithmic scale for resistance and $1/T$
for temperature.

The least squares fit assumes that the measured data will be randomly
scattered about the theoretical fit. The plot in Exercise 3.6.1 does not show
this clearly. A quick visual test of this assumption is to make a plot of the
difference between the data and the fit ie, plot the errors E_i. These are the
residuals.

Exercise 3.6.2   Plot of residuals

Make a plot of the difference between the measured data and the
theoretical fit for the data of Exercise 3.6.1. By inspection determine
if the assumption of random errors is justified.

3.7 Data modeling

The purpose of data modeling is to obtain a mathematical model
which represents a set of experimental data. First a model is chosen either on
the basis of a theory of the physical process or by guessing the mathematical
form which approximates the data. The model will have some parameters
which can be adjusted to give a best fit. For example, the model may be a
straight line $y = mx + b$ with the slope m and y intercept b as parameters.
These can be varied so that the line fits a set of data points.

Many times a model can be fitted to data to sufficient accuracy by hand
plotting. The best fit is then subjective to some degree. More accurate
determinations of model parameters can be obtained mathematically and
computationally. The first step is to form a mathematical estimate of how
well the model fits the data. One common measure of the total error in the
fit is the sum of the squares of the difference between the y value predicted
by the model, y_i^{model} and the y data value, y_i^{data}

$$\text{Total error} = e_2 = \sum_{i=1}^{N} (y_i^{\text{model}} - y_i^{\text{data}})^2 \tag{3.7.1}$$

where N is the number of data points. The difference is squared so that it is
always positive. A negative error (point above the curve) adds as much to the
total error as an equal positive error. Another measure of the error which is
sometimes used is the sum of the absolute values of the difference:

$$e_1 = \sum_{i=1}^{N} |y_i^{\text{model}} - y_i^{\text{data}}| \tag{3.7.2}$$

The total error can be calculated for a set of model parameters. The best
fit will be that set which leads to the smallest total error. A brute force way

to find the smallest error is to calculate the total error for a wide variety of parameters. The search can be narrowed to smaller parameter variations as the minimum is approached.

This method is sometimes the only possible way to proceed. However for many models, the minimum error can be found by mathematically rather than computationally varying the parameters. Since the model is a function of the parameters $y_i^{model} = f(p_1, \ldots, p_q; x_i)$ so is the error $e = f(p_1, \ldots, p_q; x_i, y_i^{data})$. The minimum of a function of a variable is found by solving the equation given by differentiating the function and setting the result equal to zero. In this case the minimum with respect to changes in the parameters is wanted so q equations are formed:

$$\partial e/\partial p_1 = 0; \ \partial e/\partial p_2 = 0; \ \ldots; \ \partial e/\partial p_q = 0$$

These can then be solved simultaneously for the parameters $p_1, p_2, \ldots, p_q$ that give the minimum. (Here it is assumed there is only one minimum and no maximum as is generally the case for physically real models.)

For example: consider the case of the simplest one parameter model, $y = p_1$; that is, the data can be represented by a constant. The total error is

$$e_2 = \sum_{i=1}^{N} (p_1 - y_i)^2$$

With its derivative

$$\partial e_2/\partial p_1 = \sum_{i=1}^{N} 2(p_1 - y_i) = 0$$

So

$$p_1 = \sum y_i / \sum 1 = \sum y_i / N$$

where the sums go from 1 to N. This is just the average value (mean) of the y data.

As a second example: consider the case where the data is to be fitted to a straight line $y = p_2 + p_1 x$ where p_1 is the slope and p_2 the y intercept. Then

$$e_2 = \sum (p_1 x_i + p_2 - y_i)^2$$
$$\partial e_2/\partial p_1 = p_1 \sum x_i^2 + p_2 \sum x_i - \sum x_i y_i = 0$$
$$\partial e_2/\partial p_2 = p_1 \sum x_i + p_2 \sum 1 - \sum y_i = 0$$

These equations can be solved directly or by forming a matrix representation

$$\begin{pmatrix} \sum x_i^2 & \sum x_i \\ \sum x_i & \sum 1 \end{pmatrix} \begin{pmatrix} p_1 \\ p_2 \end{pmatrix} = \begin{pmatrix} \sum x_i y_i \\ \sum y_i \end{pmatrix}$$

and using Cramer's rule from linear algebra to obtain the solution

$$D = N \sum x_i^2 - \left(\sum x_i \right)^2$$
$$p_1 = \left(N \sum x_i y_i - \sum x_i \sum y_i \right) / D$$
$$p_2 = \left(\sum x_i^2 \sum y_i - \sum x_i \sum x_i y_i \right) / D$$

which are equivalent to Equations (3.6.5). A three-parameter polynomial fit $y = p_1 + p_2 x + p_3 x^2$ (parabolic) can be treated in the same way.

As a final example: consider again a one parameter fit to the data but this time use the absolute value total error

$$e_1 = \sum_{i=1}^{N} |p_1 - y_i|$$

Then

$$\partial e_1 / \partial p_1 = \sum_{i=1}^{N} \text{sgn}(p_1 - y_i) = 0$$

where

$$\text{sgn}(x) = \begin{cases} 1 & x > 0 \\ 0 & x = 0 \\ -1 & x < 0 \end{cases}$$

This means that p_1 is adjusted to balance the number of y values which are greater than p_1 with the number less than p_1 (the number of $+1$s and -1s must be equal to make the sum zero). Thus, $p_1 = \text{median}(y_i \ldots y_N)$. The median can be calculated by sorting the y_i; then, $p_1 = -y_{N/2}$. In some situations the median is a better average than the mean value. If an experiment took two days to produce one number and after six days these numbers came out to be 42, 33 and 377, would you believe the mean value of 151 or the median of 42?

The model equations used in calculating the total error need not be as simple as those considered so far. An example would be the resistance variation with temperature of a thermistor $R = R_0 \exp(T_0/T)$. In this case by taking the logarithm and a change of variables, it can be expressed as a linear fit:

$$y = ax + b$$

where

$$y = \ln(R), \qquad a = T_0, \qquad x = 1/T, \qquad b = \ln(R_0)$$

In general, the model equation may not linearize. For example the expression for the heat flow in a rod $T = T_1(t_1/t)^{1/2} \exp(t_1/t)$ has this characteristic. You must start from the error expression, differentiate and solve the equations.

For some expressions not even this is possible; the trial and error method can be used. However, by using the computational speed of the computer, there is a more elegant way of searching for the best parameters. The Simplex algorithm is an iterative procedure which systematically explores the parameter values of the model. It has the virtue that any computable function can be used as a model and that no derivatives are needed. Another method commonly used is the Levenberg–Marquardt method which requires the use of error function derivatives. A good description of these algorithms can be found in the References.

3.8 Errors in data and parameters

In fitting data to a theoretical model in the least squares method used in Section 3.6, the implicit assumption has been made that each data point has been measured with the same reliability. This is often not the case and it is then important to include a measure of the data reliability when fitting a model to these data. Another result of frequent interest which is not obtainable by the simple least squares fit is to determine how much the fitted parameters can vary without straining the fit to the data (how good is the fit?).

To make a statement of how good a measurement is we usually quote the value measured together with an expected error; for example a voltage is $V \pm \Delta V$ volts. An accepted definition of ΔV is that it is the root mean square (rms) value of the random error inherent in the measurement.

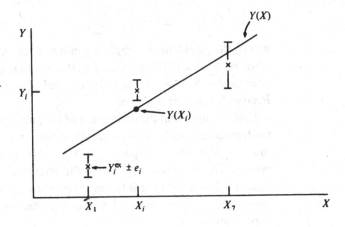

Fig. 3.9. Plot of experimental data together with a theortical fit and errors bars inherent to each data point: Y^{ex}–experimental data points, e_i–error in experimental data points, $Y(X)$–proposed theoretical fit.

Consider the plot of experimental data and of a proposed theoretical fit indicated in Figure 3.9. It shows the results of a series of measurements which yield the values $Y_i \pm e_i$ at a series of parameter values X_i. Assume the X_i are well determined. The true variation of $Y(X)$ is given as some function of X. For sake of discussion assume that Y is of the form $Y(X) = AX + B$ where the parameters A and B are to be determined.

The total error can now be written as

$$E_T = \sum_i \left[\frac{(AX_i - B) - Y_i^{ex}}{e_i} \right]^2 \tag{3.8.1}$$

where e_i is the error in the data point Y_i. A small error e_i at data point Y_i will cause the difference between the model and the data point to be weighted heavily in the sum. Thus the points with small errors have a stronger effect on the fit. Proceeding as in Section 3.6 gives the same formula for A and B (Equations (3.6.5)) except that now

$$S_{XX} = \sum_i X_i^2/e_i^2$$

$$S_X = \sum_i X_i/e_i^2$$

$$S_{XY} = \sum_i X_i Y_i^{ex}/e_i^2 \qquad\qquad (3.8.2)$$

$$S_Y = \sum_i Y_i^{ex}/e_i^2$$

$$S = \sum_i 1/e_i^2$$

By means of error propagation analysis, the errors in the estimates of A and B are determined to be

$$e_A^2 = S/D$$
$$e_B^2 = S_{XX}/D \qquad\qquad (3.8.3)$$

Where $D = SS_{XX} - S_X^2$ as before (Equation (3.6.5)). The goodness of fit of the data to the model can also be calculated:

$$G = 1 - P\left(\frac{N-2}{2}, \frac{E_T}{2}\right) \qquad\qquad (3.8.4)$$

where $P(a, x)$ is the incomplete gamma function which is tabulated in most statistics books. If G is greater than 0.1, the fit is good; if less than 0.001 then your model does not fit the data very well. Please see Press *et al*. *Numerical Recipes* for further information.

Keep in mind that the estimation of the parameters $A \pm e_A$ and $B \pm e_B$ by the least squares method is a statistical one, ie, given the data and the model junction, the calculated parameters A and B are the most likely ones for the system. The method assumes that the errors made in the measurements are random. It does not consider any systematic errors which may be lurking in your data. These last need to be ferreted out by careful thought and experimentation.

Exercise 3.8.1 Errors in thermistor data

Make an evaluation of the error in your resistance determinations with the ADC and reanalyze the thermistor data with error considerations. To simplify the error analysis, assume some reasonable constant error ($\Delta R_i = \Delta R$ for all i) and simplify the error equations by factoring the error out of the sums.

3.9 Digital signal processing

Proper use of the ADC requires analog signal conditioning before the ADC samples the data as described in Section 3.2. Once in the computer, a series of samples can then be analyzed to emphasize various features of the data.

If the data has some noise mixed in with a broad trend, a smoothing

Fig. 3.10. Data smoothing example.

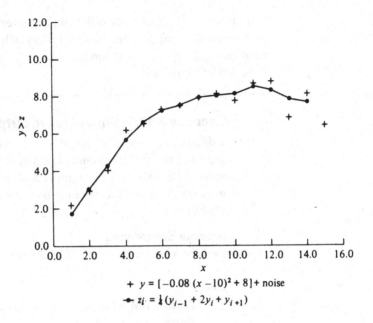

$$+ \quad y = [-0.08 (x - 10)^2 + 8] + \text{noise}$$
$$\bullet \quad z_i = \tfrac{1}{4}(y_{i-1} + 2y_i + y_{i+1})$$

process can be used to suppress the noise. One common method is to apply an averaging scheme as shown in Figure 3.10. The new point z_i is a weighted average of the old point y_i with its neighbors. Specifically,

$$z_i = \tfrac{1}{4}(y_{i-1} + 2y_i + y_{i+1}) \tag{3.9.1}$$

Equation (3.9.1) can be extended to more points if more smoothing is required.

This way of smoothing data is one example of a digital low-pass filter; it suppresses the high frequency components of the time series. Another way of making a digital filter is by the recursive procedure:

$$z_i = (1 - \alpha)y_i + \alpha z_{i+1} \tag{3.9.2}$$

When applied to a time series, Equation (3.9.2) approximates a low-pass analog resistor–capacitor filter with the parameter α setting the frequency cut-off. Again, more smoothing can be done by including more terms in the recursion. Both recursive and non-recursive filters can be constructed which will act as high-pass filters if the interesting part of the signal is not the trend but the time varying part.

A reminder: digital filtering in no way replaces analog filtering before sampling the signal. Aliasing occurs when the data is sampled and cannot be remedied later.

Much more elaborate signal processing is often required to analyze a set of data. The references contain further information.

3.10 Controlling computer output ports

The next task you will work on is to use the thermistor in a temperature controller. The computer will not only be measuring the temperature of the block but will turn the heater on and off to maintain the predetermined temperature. To get a feeling of how to send output from the computer you will use a program to generate square waves at an output port of the 8255 PPI interface.

Exercise 3.10.1 Square wave output

Connect the oscilloscope to the terminal marked PA0 on the protoboard. This is channel of port A of the digital interface. Write and run the following program and adjust the oscilloscope until you get a steady trace. Measure the time the output is high and the time which it is low.

program SquareWave;
{outputs a square wave on PA0 of the 8255 as fast as it can}

const
      PortControl = 803;   {the control port for 8255 #1}
      PortA   = 800;   {the register for port A #1}

begin
      Port[PortControl] : = 128;   {set port A as output in mode 0}
                                     {ports B and C are also outputs}

     repeat
      Port[PortA] : = 1;   {set PA0 high, keep PA1–PA7 low}
      Port[PortA] : = 0;   {set PA0 low, keep PA1–PA7 low}
     until false;   {use 'CTRL C' to stop program}
   end.

On the same board as the ADC inside the PC is an 8255 PPI chip. Its purpose is to allow digital data to be output from or input to the computer. As with the ADC which you used before, the control of the digital I/O interface is done by addressing and sending data to registers of the chips. Appendix E contains the data sheets which describe the details of its operation.

Each 8255 has a control register and three data registers to control ports labeled A, B and C. The control register instructs the ports to be either input or output. If a port is an input port the data in the corresponding data register will reflect the digital voltages on the port wires. If a port is an output then the port wires will have digital voltages which are controlled by the data in the data register.

The program in Exercise 3.10.1 writes the number 128 (binary 1000 0000) to the control port. This sets all the ports as outputs. Then the number 1 (binary 0000 0001) is sent to the port A data register which turns on the wire corresponding to PA0 while leaving the other wires (PA1–PA7) off. The number 0 (binary 0000 0000) then turns off PA0.

Exercise 3.10.2 Square wave output on PA3

Modify and run the program of Exercise 3.10.1 so that it will generate square waves on wire PA3. Observe with the oscilloscope.

3.11 Port and Mem statements

Port and Mem statements are used to address registers and memory locations in the PC. Port is used for the special I/O addresses and Mem for regular addresses. To address a word (two sequential bytes or registers) instead of a byte the statements PortW and MemW are used. The statements

Port[Address] : = Data;

and

Data : = Port[Address];

are conjugate statements, ie, the first will place the byte Data at the I/O location Address and the second will obtain the byte from the location Address and put it in the variable Data.

The Mem statement works the same way except that the specification of the address needs two parts: the Segment and the Offset. Thus, the statements become

Mem[Segment : Offset] : = Data;

and

Data := Mem[Segment : Offset];

The Segment and the Offset are combined to give the address. Details of how this is done will be discussed in Chapter 6.

3.12 Using a HEXFET to control the heater

The digital signals coming out of the PC are feeble and in general cannot drive external circuitry loads directly. HEXFETs are one variety of enhanced mode power FETs (Field Effect Transistors) which are particularly suited for controlling large amounts of power by using the digital control signals coming out of a computer.

To see how these devices are used, set up the circuit as shown in Figure 3.11. It will act like the push button switch you used earlier but will be controlled by the computer. When a HI signal is applied to the gate of a HEXFET, the device conducts current like a closed switch; when a LO signal is applied, the device acts like an open switch, ie, it has infinite resistance.

Fig. 3.11. HEXFET connections
and pin diagram for thermistor
apparatus.

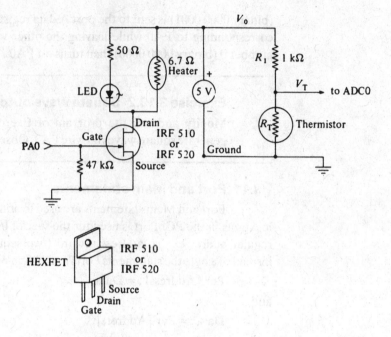

Exercise 3.12.1 Controlling the HEXFET

Connect the gate of the HEXFET to PA0 but wait to turn on the
power. Use the following program to test the control of the HEX-
FET. Turn on the power to the HEXFET after the program is
running so that the HEXFET is not inadvertently left on for a long
period of time and the thermometer is overheated.

```
program TestHEXFET;
{to test the on/off operation of the HEXFET}

uses crt;                              {this statement is necessary because}
                                       {the crt unit contains the procedure}
                                       {delay(n : integer); used below}

const
     PortControl = 803;                {the control port for 8255 #1}
     PortA       = 800;                {the register for port A #1}

begin
     Port[PortControl] := 128;         {set port A as output in mode 0}
                                       {ports B and C are also outputs}
```

```
repeat
    Port[PortA] : = 0;              {init PA0 low}
    Delay(250);                    {pause for a moment}
    Port[PortA] : = 1;             {set PA0 high}
    Delay(250);                    {pause}
until keypressed;
    Port[PortA] : = 0;             {leave program with HEXFET off}
end.
```

Exercise 3.12.2 Temperature controller

Write a program for a temperature controller following the flow chart in Figure 3.12. The program should ask you to type in a temperature. The computer should then turn the power to the heater on and off in response to the thermistor voltages read. Run the program and demonstrate to your laboratory instructor that the thermometer does stabilize to the temperature typed in. When testing be sure to turn off the heater manually or with a program statement at the end of your program to ensure that the heater doesn't overheat.

Fig. 3.12. Flow chart for temperature controller.

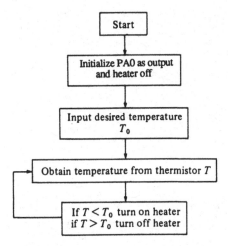

Exercise 3.12.3 Temperature controller with hysteresis

So that heater is not turning on and off rapidly at the desired temperature, modify the program to turn on the heater when the temperature is below desired temperature minus $1.0° (T_0 - H)$ and turn off the heater when it is above the desired temperature plus $1.0°$ $(T_0 + H)$.

The process of turning the heater on and off used in the program of Exercise 3.12.3 is called hysteresis. It is used in many process control situations to stabilize the system. A thermostat for a household furnace uses hysteresis so that the furnace doesn't turn on and off too quickly. In a later section you will be using a Schmitt trigger which uses hysteresis to stabilize voltages.

4 Timing

In many experiments, the measurement of interest is the change with time of a particular quantity (eg, dx/dt). One of the most useful capabilities of a computer is to provide accurate and varied timing signals so that these measurements can be made. Indeed, the internal operation of the computer requires the orchestration of many events to the beat of the internal clock. In this section several ways of generating time intervals will be presented.

4.1 Timing loops

A simple method of generating time intervals is to use the time required by the computer to execute instructions. This method is difficult to make precise but nevertheless is useful in some situations.

Exercise 4.1.1 Timing loops

(*a*) Retrieve the program SquareWave which uses PA0 as an output. With the program running, use the oscilloscope to observe the output. Again note the time it takes for one period and the time PA0 is high and the time it is low.

(*b*) Try adding statements between the two Port statements and determine how long they take. Some interesting ones are:

```
j : = i * i + 1;          {be sure to declare var i,j : integer}
x : = i * i + 1.0;        {with var x : real}
x : = cos(pi * i);        {with var x : real}
write('A');               {write a character to the screen—
for i : = 1 to 10 do;     {an empty loop executed 10 times}
```

(*c*) The Turbo Pascal statement Delay(n); provides an easy way to pause the execution of a program. Insert this statement between the Port statements and vary n to determine the time the delay takes.

This type of timing loop could be placed anywhere in a program to provide a delay. However, it suffers from several disadvantages. Since every statement takes a different amount of time, it is difficult to predict the exact amount of time a loop will take. Trial and error and an oscilloscope must be used to obtain a particular desired time. Also, only a delay in program execution is really possible. Measurements of time intervals between different parts of a program cannot be done.

The on board 8253 or 8254 timer chip and the interrupt clock which will be discussed later provide a way to do timing which is fast, accurate, and independent of the program statements. Never-the-less, the Delay(n) statement can be useful; the next section illustrates one use.

4.2 Stepping motors

Stepping motors are used to position apparatus of all kinds precisely. A stepping motor rotates a shaft a small increment of a turn for each pulse of electric current it receives. An electric clock is a stepping motor: it rotates a fixed, small amount for every pulse of current it receives from the wall power outlet. The power outlet provides the current which reverses polarity 60 times each second so that by using gears the hands rotate at the proper speed. A clock motor always rotates in a fixed direction (unidirectional). Some stepping motors can be made to rotate either clockwise or counter-clockwise under computer program control (bidirectional); the one which you will use and the one in the disk drive which positions the reading head are bidirectional.

Exercise 4.2.1 Single step of stepping motor

(a) To see how a stepping motor and controller IC (Integrated Circuit) are used, make connections to PA0 and PA1 as indicated in Figure 4.1. Write a procedure which generates a single negatively going pulse on PA0; ie, PA0 should be normally high, then go low briefly, then return high. Write a program to test the procedure to see that each procedure call causes the motor to step once. Depending on the particular stepping motor used, the time at which the pulse is low may need to be lengthened by use of a Delay(n) statement.

Fig. 4.1. Stepping motor controller (4202) connections.

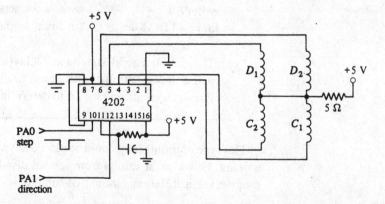

(b) Write a second procedure which, in addition to giving a proper pulse, allows control of the direction of rotation of the motor by specifying the level of PA1. The awkward voltage programming for the motor itself is done by the 4202 control IC so that in order to step the motor you need only specify the direction by setting the polarity of the direction control wire and then apply a high, low, high pulse to the stepping input. Be sure that the direction wire keeps a constant voltage as the pulse is generated.

The mechanics of a stepping motor are shown in Figure 4.2. The rotor is a permanent magnet with 12 sets of north and south poles; the stators each have 12 sets of fingers which can be magnetized electrically. Each stator has 2 coils of wire inside it, labeled C and D. If coil C is energized the fingers marked A become north poles, those marked B become south poles. Coil D energizes the stator with reversed current direction so that each A becomes a south pole and each B a north.

Figure 4.2(a) shows the motor pulled apart to show the relationship of the stators and the rotor. Figure 4.2(b) shows the rotor unravelled with its north and south poles lying next to one another. There are two sets of stators, the fingers of each are displaced from one another in azimuth as shown in Figure 4.2(a). With the rotor unravelled as shown, it can be pulled to the right by energizing the A2B2 stator set with coil D2 so that A2 becomes a south pole and B2 a north pole. The rotor will then move over one step so that the north poles of the rotor lie under the south poles of the stator 2.

Fig. 4.2. Stepping motor opened up. (a) Stators 1 and 2 each have 12 pairs of poles (A1, B1, A2, B2) and 2 coils (C1, D1, C2, D2). Current in coil C makes A poles north and B poles south and current in coil D makes A poles south and B poles north. (b) Poles and rotor flattened out to show staggering of stator 1 and stator 2's poles. To move rotor one position to the right from the position shown, turn coil D1 off and D2 on (A2 is then south and B2 north). To move once to left, turn D1 off and C2 on (A2 north and B2 south).

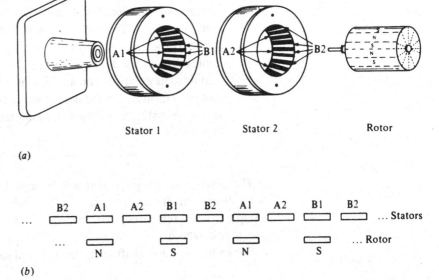

(a)

(b)

The controller IC has two inputs: a direction control and a step control. A high level on the direction control signals movement in one direction, a low level in the other. The controller is set so that each time the voltage goes from low to high on the step control input, the stepping motor will advance two steps in the appropriate direction. Between pulses the step control should be left high. In the case of the stepping motor illustrated, one step is $360/(4 \times 12) = 7.5°$ since it has 12 poles and each step moves the rotor one fourth of a pole distance. Thus one pulse on the step control line will move the shaft 15°. The controller IC regulates the current flow in the four windings of the stator. It has logic circuitry within it so that it knows which coil must be energized to step in the specified direction from where it is. This saves you the trouble of programming these details. If the stepping motor shaft is connected to a gearbox with a 200:1 gear ratio, the output shaft will turn one revolution for every 200 revolutions of the motor shaft (200:1 gear ratio).

Exercise 4.2.2 Maximum stepping rate

(a) You have seen that it is only necessary to use two Port[ ]; statements which make PA0 go low and then back to high to step the motor once. The stepping motor is a mechanical device which is inherently slow; thus it is important that there be a reasonable amount of time between pulses. By using a Delay(n) statement to waste time between pulses, determine the maximum number of pulses per second that the motor will respond to. Do this by varying the time delay and watching to see whether the motor responds properly or not. For example, a program which gives 200 pulses to the stepping motor of Figure 4.2 with the gearbox attached should turn the gearbox output shaft 15°. Use the oscilloscope to measure the time between pulses. Test your motor in both directions.

(b) Write a procedure, to be used by later programs, which will step the motor once in the direction specified before entering. The program should delay the proper amount of time for the stepping motor before returning. Set the delay time so that you never ask the motor to rotate faster than $\frac{1}{2}$ the maximum rate. This will ensure reliable operation.

The positioning program you will be asked to write next is subject to truncation errors if care is not taken. Truncation error is always present in a computer using real arithmetic. It also results from the finite step length of the stepping motor.

For example, if the shaft is to turn 15°, 200 steps are given to the motor.

However, if the shaft is to turn $10°$ then $200(10/15) = 133\frac{1}{3}$ steps should be given. Since the motor cannot turn $\frac{1}{3}$ of a step, 133 steps is the closest which is possible. If the angle is stored in a variable as $10°$ and not $15(133/200) = 9.98°$ then $0.02°$ error is made. This may seem small but these errors can accumulate over many shaft turns to yield a large error.

Conversion of a number to the internal representation in a computer also creates truncation error. All numbers are converted to binary form when entered into a computer. This conversion is good only to a specified number of bits so that numbers like $\frac{1}{3}$ are only approximated. Like the stepping motor, binary notation has only a finite number of steps to represent a number. If the desired number falls between two of the steps, the closest representation is used. It is also like the ADC which has a resolution of a specified number of bits.

Exercise 4.2.3 Truncation errors

(a) Pick up a hand calculator and calculate the following number: $x = \frac{4}{3} - 1$. Now calculate $4x - 1$, and repeat many times. Watch the truncation error grow.

(b) Enter and run the following program to see how well this goes in the computer.

```
program Truncerr;
{to show the limitation of arithmetic on computers}

var i : integer;
    x : real;

begin
  x : = 4.0 / 3.0 - 1.0;
  for i : = 1 to 30 do begin
    writeln( i,x);
    x : = x * 4.0 - 1.0;
  end;
end.
```

Exercise 4.2.4 Positioner

Using the procedure in Exercise 4.2.2 write a program which moves the output shaft of the gearbox to a specified angle. At the outset the current position of the stepping motor should be read into the program. The program should then ask you for the angular position of interest and step the motor to that angle. After it is at this position the program should come back and ask for the next desired position.

To avoid truncation errors in calculating the angle, keep track of steps not degrees. The program should accept positive or negative angles of any magnitude and turn the shortest route to the angle.

4.3 Number systems

To work with I/O devices and with assembly language programs, it is necessary to go back and forth among the representations of numbers in decimal, hexadecimal, and binary. In Turbo Pascal integer numbers can be written in decimal or hexadecimal notation. However, internally the computer represents all numbers and characters in binary (base 2). This internal conversion to binary is usually not important to the user but becomes so when connecting I/O devices to the computer. Then a binary representation directly corresponds to signal levels on the I/O lines. Hexadecimal numbers (base 16) are a convenient shorthand notation for long binary numbers.

When a decimal number is written down, say 348, what is really indicated is that there are 3 hundreds, 4 tens and 8 ones (Figure 4.3). This can be described by the equation

$$348 = 3 \times 10^2 + 4 \times 10^1 + 8 \times 10^0 \tag{4.3.1}$$

Fig. 4.3. Decimal, binary, and hexadecimal number representations.

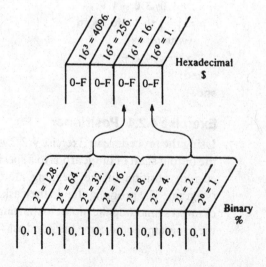

Table 4.1 *Correspondence between binary, hexadecimal, and decimal characters*

Binary %	Hexadecimal $	Decimal
0000	0	0
0001	1	1
0010	2	2
0011	3	3
0100	4	4
0101	5	5
0110	6	6
0111	7	7
1000	8	8
1001	9	9
1010	A	10
1011	B	11
1100	C	12
1101	D	13
1110	E	14
1111	F	15

In exactly the same spirit a hexadecimal number with the characters 1234, represents

$$\$1234 = 1 \times 16^3 + 2 \times 16^2 + 3 \times 16^1 + 4 \times 16^0 \qquad (4.3.2)$$

It is useful to remember that $16^3 = 4096$, $16^2 = 256$, $16^1 = 16$, and $16^0 = 1$. Here, as in Turbo Pascal, hexadecimal numbers are indicated by a $ sign preceding the number. Sometimes a period is used to indicate a decimal number even if it is an integer (for example 348.). Each of the characters 1, 2, 3, 4 in Equation (4.3.2) could be a number from 0 to 15 just as in the decimal representation each place (column) has a number between 0 and 9 (Figure 4.3). In hexadecimal, to represent 10, A is used, 11, B, etc., as shown in Table 4.1. As an example 348. = $15C = $1 \times 256 + 5 \times 16 + 12 \times 1$.

A number will be preceded by a % sign to indicate that the characters which follow are a number in binary representation. Thus

$$\%0101 = 0 \times 2^3 + 1 \times 2^2 + 0 \times 2^1 + 1 \times 2^0 \qquad (4.3.3)$$

The magic numbers here are $2^7 = 128$, $2^6 = 64$, $2^5 = 32$, $2^4 = 16$, $2^3 = 8$, $2^2 = 4$, $2^1 = 2$, $2^0 = 1$. In writing down binary numbers it is convenient to write them down in groups, four digits (bits) at a time. This makes it easy to identify the position in which each bit belongs. It also makes it easy to go back and forth between binary and hexadecimal since 4 binary bits = 1 hexadecimal character. Thus, 348. = $15C = %0001 0101 1100.

To allow you to get a feeling for the internal representation of numbers, eight Light Emitting Diodes (LEDs) have been connected to the eight wires of port A of a second 8255 PPI chip. A wiring diagram is shown in Figure 4.4. When the port is set as an output and a 1 is put into a particular bit of the

Fig. 4.4. LED number display wiring. Schematic of LEDs on port A. The 74LS04 IC is a BUFFER/DRIVER to provide current drive for LEDs. A 'high' on port A will illuminate an LED.

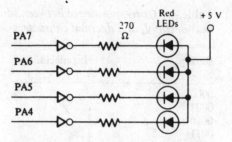

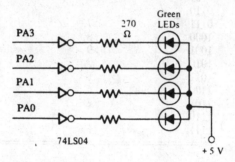

register, the corresponding LED will light up. Data lines 0–3 are connected to green LEDs and 4–7 to red ones. The 74LS04 IC between the actual port wires and the LEDs has what are called line drivers or buffers to provide the current required to light up the LEDs. The ports alone are not capable of generating enough power.

Exercise 4.3.1 LED binary number display

(a) Write a program which sets the data wires of port A of 8255 PPI #2 as output and sends various numbers to the port. (The control register is at I/O address 807 and the port A address is at 804.) Then:
  (i)   Send data so that wires 3, 5 and 7 are high and the rest are low.
  (ii) Send the numbers $28 and $C3 and verify that the expected LEDs light up.
  (iii) Send the decimal or hexadecimal equivalent of %1010 0111 to the port and verify.

(b) Even though the port is set up as an output, it is possible to read the data in the register by using the Port statement. Store and read several numbers in the port to verify that this can be done.

(c) To see how binary counting works, run the following program:

program BinaryCount;
{outputs a binary counting sequence to port A of 8255 PPI #2}
{for visualization with the LEDs}

```
uses crt;                              {for Delay procedure}

const
        PortControl = 807;             {the control port for 8255 #2}
        PortA       = 804;             {the register for port A #2}

var
        i : integer;                   {a counter}

begin
    Port[PortControl] : = 128;         {set port A as output in mode 0}
                                       {ports B and C are also outputs}


    for i : = 0 to 255 do begin
        Port[PortA] : = i;             {send data to port}
        Delay(500);                    {wait for a while}
    end;
    for i : = 255 downto 0 do begin
        Port[PortA] : = i;
        Delay(500);
    end;
end.
```

The representation of negative integer numbers in a computer is done somewhat differently than in common computations. In many ways the system is more logical than the customary one, and it certainly makes things much simpler in a computer. Imagine a continually incrementing binary four-bit counter. A counter is a device which increments each time a pulse is applied. Table 4.2 shows the sequence of digits as the count pulses are added (start at 0 and read upwards). The correspondence to the ordinary number system is shown in the column to the right. Being a four-bit counter the 'readings' repeat every 16 counts, ie, the count after 1111 gives 0000. (This transition is called overflow.) A representation of the numbers 0–15 is naturally done through the correspondence to the counter readings of 0000–1111. One possible way to represent positive *and* negative numbers is to assign the numbers between −8 and 7 to the counter readings of 1000–0111. This is called the two's complement representation. As shown in Table 4.2, −1 becomes 1111, −2 becomes 1110, etc. Notice that the most lefthand bit takes on a special meaning; if it is 1, a negative number is being represented, if 0, a positive one. Within the computer a particular bit combination, say %1011 will represent 11 one time and −5 at another.

Also note that the four-bit counter will repeat its reading every 16 counts. Thus the decimal numbers 11, 27, 43, etc., will all be represented by the same

Table 4.2 *A four-bit counter*

Four-bit binary	Negative numbers interpretation	No-negative numbers interpretation
1111	−1	15
1110	−2	14
1101	−3	13
1100	−4	12
1011	−5	11
1010	−6	10
1001	−7	9
1000	−8	8
0111	7	7
0110	6	6
0101	5	5
0100	4	4
0011	3	3
0010	2	2
0001	1	1
0000	0	0
1111	−1	15
1110	−2	14
1101	−3	13
1100	−4	12
1011	−5	11
1010	−6	10
1001	−7	9
1000	−8	8

Add

Subtract

binary combination in a four-bit counter. If we count down 16 counts from 11 we come back to the same binary reading; thus 11 and −5 (which is 11−16) are repesented by the same binary string. Similarly in a 16-bit counter, the bits will repeat every 65 536 counts (2^{16} = 65 536). By the same reasoning the number 38 060 and the number −27 476 will both be represented by the same 16-bit string since 38 060 − 65 536 = −27 476.

4.4 Using the internal clock

In order to determine time intervals between events some sort of clock needs to run over the interval. One which runs independently of other operations of the computer provides the most flexibility. The computer can then just look at the clock when it needs to and ignore it otherwise.

The IBM-PC keeps track of the time by means of a counter which increments approximately every 55 ms (actually 54.925 ms), ie, it has a clock with tocks (ticks come later) every 55 ms. The counter is four bytes long so that up to a full day (1 573 040 or $1800B0) of counts can be stored. The counter bytes are at memory locations $46C–$46F of segment $0000. The current value of the counter can be read by means of the Mem or MemW statements. The DOS operating system reads this counter and calculates the hours, minutes, and seconds for the Time DOS command.

If intervals shorter than a day are needed, only part of the counter may be used. For example, reading the lowest two bytes (at $46C and $46D) could time intervals up to an hour long.

Exercise 4.4.1 Reading the internal clock

Write a procedure using the MemW statement which will read and return the lowest two bytes of the TimeOfDay counter. Test the procedure with a program which gets the number of tocks and prints it on the screen. Add statements to print out the time in seconds. Run each program several times to see that the value does change.

Exercise 4.4.2 A stopwatch

To measure time intervals with the computer is now just a matter of reading the internal clock twice and subtracting the times. Write a program which will print out the elapsed time from the start of the program each time a key is pressed on the keyboard. Also print out the time interval from the last time a key was pressed. Make sure that your program compensates for the case when the second time is less than the first; ie, when the counter has counted beyond its maximum count and is starting over (the counter has rolled-over).

Exercise 4.4.3 A beeper

Write a program which beeps the terminal bell at one second intervals. Use the statement Write(CNTRL-G) to ring the bell.

---

5 Thermal diffusion

The experiments which you will be called upon to do in this chapter give you a chance to apply the timing concepts of Chapter 4 and to review the use of the ADC while learning about the phenomenon of diffusion. Specifically, you will be studying thermal diffusion but many of the concepts encompass a variety of other phenomena.

5.1 Heat flow equation

In this section you will explore some of the physical and mathematical considerations of one-dimensional heat diffusion. When heat is added to a material there are two parameters which affect the distribution of temperatures: the specific heat (or heat capacity) and the thermal conductivity. The specific heat indicates how much heat is added to a mass of material for a specified temperature rise. The thermal conductivity indicates how fast the thermal energy is transported through the material.

Consider the flow of heat in a rod as shown in Figure 5.1. The specific heat C of a material is the ratio of the amount of heat added dq (joules) to the resulting rise in temperature dT (degrees Kelvin) per unit mass dm (kg); thus $C = (dq/dT)/dm$, (see Equation (3.3.1)). For a rod of cross-sectional area A, the volume $dV = A\,dz$ and $dm = \rho\,dV$ where ρ is the density. So, the amount of heat added to the length dz of the rod is

$$dq = C\rho A\,dT\,dz = sA\,dT\,dz \qquad (5.1.1)$$

where s is the volumetric heat capacity, $C\rho$.

When one end of the rod is hotter than the other there will be a net flow of energy from the hot end to the cool end. The power P (watts) of this heat flow down the rod is the heat energy per unit time flowing past a point on the rod $P = dq/dt$ (Equation (3.3.2)). For one-dimensional heat flow, P is proportional to the temperature gradient dT/dz, the thermal conductivity k (W/m K) and the cross-sectional area;

$$P = -kA\,(dT/dz) \qquad (5.1.2)$$

There is a minus sign because heat flows from higher to lower temperatures. In writing this equation, it is assumed that the rod is insulated; no heat escapes from the rod by conduction, convection or radiation. The net heat gain per unit time dq/dt in the piece of rod between z and $z + dz$ is given by the difference in the power flowing in at z and the power flowing out at $z + dz$, so

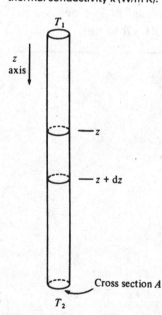

Fig. 5.1. The flow of heat in a rod of specific heat C (J/kg K) and thermal conductivity k (W/m K).

$$dq/dt = P(z) - P(z + dz) = -(\partial P/\partial z)dz \qquad (5.1.3)$$

Combining Equations (5.1.1), (5.1.2) and (5.1.3) gives the differential equation for heat flow in a rod

$$s(\partial T/\partial t) = k(\partial^2 T/\partial z^2) \qquad (5.1.4)$$

This equation has many solutions; if a quantity of heat is added to the rod quickly (a heat pulse), the solution can be written as follows:

$$\left.\begin{array}{l} B_1 = \text{constant} \\ B_2 = \text{constant} \\ T(t, z) = B_1 + B_2 \exp(-z^2 s/4kt)/t^{1/2} \end{array}\right\} \qquad (5.1.5)$$

Details of how this solution can be obtained are found in Appendix F.

Exercise 5.1.1 Impulse heat diffusion solution

(a) Show that Equation (5.1.5) is a solution of Equation (5.1.4).
(b) Show that B_1 can be interpreted as the starting temperature; that is, $T(t, z)$ at $t = 0$ for $z \neq 0$. $B_1 = T_s$.

The solution (5.1.5) describes the temperature at any point in the rod as a function of time after an impulse of heat has been added at $z = 0$. Before proceeding further it is useful to examine the graphs of temperature T vs. distance z of the solution at various times after the impulse. These are shown in Figure 5.2. In this figure, $T = T(t, z) - T_s$. At times near zero, the heat, and thus the excess temperature, is concentrated near $z = 0$. As time progresses the heat diffuses away from the center to larger and larger values of z with the peak temperature decreasing in time.

Fig. 5.2. Heat flow in a rod, temperature vs. distance where $t = t's/k$.

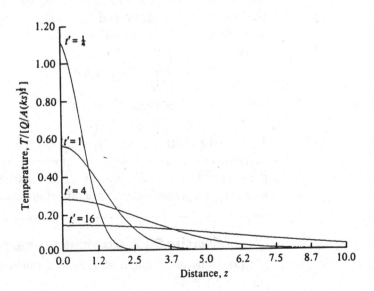

An important point is that since the solution is symmetric with respect to z, just as much heat diffuses up as down the rod. Since there is no heat flow across the cross section at $z = 0$, cutting the rod at $z = 0$ will not modify the form of the solution although now all the heat added goes one way. This half-space rod is the configuration which you will study experimentally.

To obtain a theoretical expression convenient for analyzing a quantitative experiment, it is useful to relate the constant B_2 in Equation (5.1.5) to the total heat Q added to the rod (from $z = 0$ to $z = \infty$) by integrating Equation (5.1.1). Consider T as the excess temperature above $T(0)$, ie, $T = T(t, z) - T_s = T' - T_s$; integrating Equation (5.1.1) from temperature T_s to T' gives

$$dq = sA\,dz(T' - T_s) = sAT\,dz \qquad (5.1.6)$$

To integrate from $z = 0$ to $z = \infty$, use Equation (5.1.5) to describe the variation of temperature at any z and t. Then

$$Q = \int_{z=0}^{\infty} sA \, \frac{B_2 \exp(-z^2 s/4kt)}{t^{1/2}} \, dz = \frac{sA}{t^{1/2}} B_2 \int_0^{\infty} \exp(-z^2 s/4kt)dz$$
$$= \frac{sA}{t^{1/2}} B_2 \frac{\pi^{1/2}}{2} \left(\frac{4kt}{s}\right)^{1/2} = B_2(\pi ks)^{1/2}A \qquad (5.1.7)$$

solving for B_2 and inserting into Equation (5.1.5)

$$T(t, z) = \frac{Q}{A} \frac{1}{(\pi ks)^{1/2}} \frac{\exp(-z^2 s/4kt)}{t^{1/2}} + T_s \qquad (5.1.8)$$

As written Equation (5.1.8) is not in an optimum form for displaying some of the important features it contains. It is often very helpful, particularly for purposes of recognizing the domain of behavior in a given physical situation, to relate the quantities in an equation to physically significant parameters rather than simply measuring time in seconds, temperature in degrees centigrade, etc. You saw this before in the equation for the thermistor resistance as a function of temperature of Chapter 2, the natural parameters there being R_0 and T_0. For displaying the change in temperature T as a function of time t at a fixed z, Equation (5.1.8) can be written in terms of a characteristic time t_1 and a characteristic temperature T_1 as

$$\left.\begin{array}{l} T/T_1 = (t_1/t)^{1/2} \exp(-t_1/t) \\ t_1 = sz^2/4k \\ T_1 = 2Q/Azs\pi^{1/2} \\ T = T(t, z) - T_s \end{array}\right\} \qquad (5.1.9)$$

Equations (5.1.9) immediately show several important points. First, the variation of temperature with time at a constant z can be related to just two parameters t_1 and T_1. Second, the characteristic time scale t_1 is proportional to z^2; this is a general property of diffusion phenomena.

Exercise 5.1.2 Graphing the heat diffusion equation

(a) Generate a graph of T/T_1 as a function of t/t_1 from $t/t_1 = 0.1$ to $t/t_1 = 10$.

(b) Show that the temperature T_1 is proportional to the temperature rise which a quantity of heat Q would produce if absorbed by a length z of the rod with the heat distributed over the length of the rod. Find the constant of proportionality.

(c) The temperature T_1 can also be related to the maximum values which T assumes, show that the maximum occurs at $t/t_1 = 2$ and at the maximum $T/T_1 = 0.43$.

5.2 Numerical integration of the heat flow equation

Appendix F shows a solution to the differential equation for one-dimensional heat flow for an impulse of heat at $t = 0$. For other starting conditions or parameters dependencies the equation could be much harder, if not impossible, to solve. For example, the thermal conductivity k is really temperature dependent $k = k(T)$ and so cannot be treated as a simple constant parameter. An analytical solution quickly becomes impossible and you must resort to numerical solutions.

General numerical integration of partial differential equations is a broad and difficult subject. The following will be a simple procedure which works in this case but must be used with care. It is really only meant to illustrate a general approach. For further discussion see *Numerical Recipes The Art of Scientific Computing*, by Press *et al.* in the bibliography.

The basic equations for the flow in a rod are the static equation for the heat capacity Equation (5.1.1) and the dynamic equation with the thermal conductivity Equation (5.1.2) which are combined to form the differential equation, Equation (5.1.4). However for purposes of numerical integration, it is best to leave them separate and write them in this form:

$$\Delta Q = -kA \, \Delta T \, \Delta t / \Delta z \tag{5.2.1}$$
$$\Delta T = \Delta Q / As \, \Delta z \tag{5.2.2}$$

where Δ is assumed to approach zero.

Now break up the length of the rod (Figure 5.1) into N_z pieces of length Δz each and consider the ith piece; the heat flowing into this piece in the time Δt will be:

$$Q_{in} = kA(T_{i-1} - T_i) \, \Delta t / \Delta z \tag{5.2.3}$$

If the temperature in element $i - 1$ is hotter than in the element i then Q_{in} will be positive. The heat flowing out of the piece will be:

$$Q_{out} = kA(T_i - T_{i+1}) \Delta t / \Delta z \tag{5.2.4}$$

The difference of the two is the heat gained or lost in the element:

$$\Delta Q_i = Q_{in} - Q_{out} \tag{5.2.5}$$

This heat changes the temperature of the element in proportion to its heat capacity:

$$\Delta T_i = \Delta Q_i / As \, \Delta z \tag{5.2.6}$$

and so

$$T_i^{\text{new}} = T_i^{\text{old}} + \Delta T_i. \tag{5.2.7}$$

Exercise 5.2.1 Integration algorithm

(a) Equations (5.2.3)–(5.2.6) can be used to determine the temperature in any element at any time (which is all we want out of a solution to the differential equation) as follows:

First specify Δz and N_z and the temperature T_i in each of the elements $i = 1 \ldots N_z$ at the start, which in the case of the laboratory experiment will be $T_1 = Q$ (from heater)$/As \, \Delta z$ and $T_2, T_3, \ldots,$ $T_{N_z} = 0$. Also specify the time step desired Δt.

Next, make the calculations in Equations (5.2.3)–(5.2.6) for each of the elements using the old temperatures to give new temperatures.

Repeat the last step until the desired time is reached.

Now repeat the whole procedure but with a smaller Δz and/or smaller Δt. Compare these results with the previous ones to make sure that they are not sensitive to the size of the steps used. If they are, reduce the step size again.

Since the theory deals with an infinite rod, another parameter which needs to be examined is the length of the rod $N_z \Delta z$. Make sure that it does not affect the results.

(b) With a working program in hand, the results can be checked by comparison with the analytical solution. But now the analysis can be taken further; consider the following questions and how the program might change to answer them:

What is the effect of a short rod or a rod with one end clamped at a constant temperature?

What is the effect of a thermal conductivity k which is a function of temperature, eg, $k = k'/T$?

What is the effect of the heat impulse occurring over a longer interval of time?

How does convective and radiative heat loss affect the temperature distribution?

5.3 Experimental setup and program development

The apparatus for these experiments is illustrated in Figure 5.3. In the top of the copper rod (#10 copper wire, 2.59 mm diameter) is set a 3.3 Ω resistor which is used as a heater. Current can be switched into the heater under program control using the IRF 520 HEXFET in a manner similar to that used in Section 3.12. After generating a short pulse of heat by

Fig. 5.3. Heat diffusion apparatus.

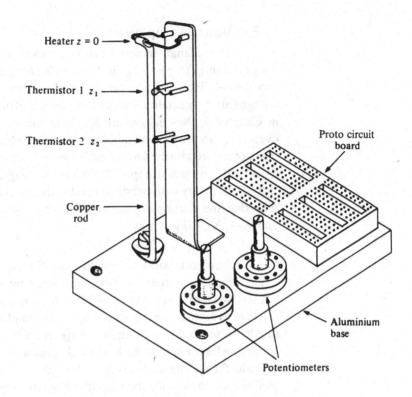

Heater $z = 0$

Thermistor 1 z_1

Thermistor 2 z_2

Proto circuit board

Copper rod

Aluminium base

Potentiometers

momentarily turning on the HEXFET, the computer will measure the increase in temperature at two positions down the rod using two thermistors. The thermistor positions are as shown on Figure 5.3. A plot of the temperature vs. time at each of these thermistors will yield values for the heat capacity and thermal conduction constants of copper and also demonstrate the functional dependence of heat diffusion on time and distance.

Fig. 5.4. Flow chart for Exercise 5.3.1.

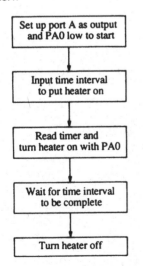

Set up port A as output and PA0 low to start

Input time interval to put heater on

Read timer and turn heater on with PA0

Wait for time interval to be complete

Turn heater off

Exercise 5.3.1 Heat impulse to rod

Write a procedure which uses the timing concepts of Chapter 4 to turn the heater on for an amount of time which you type as input data into the computer. Use a PA0 to control the HEXFET. A flow chart outlining the steps in the program is shown in Figure 5.4. Check your program and apparatus by putting an oscilloscope probe between the heater and ground and then turning the heater on for times ranging from 0.1 s to 2 s. Note the voltage across the heater when it is on with the oscilloscope and compute the power being put into the heater. (Remember: do not put the probe ground clip to any circuit point which is not at ground potential!)

5.4 Voltage amplifier

The change in temperature of each thermistor from an initial temperature $(T(t, z) - T_0)$ is the significant quantity to measure in this experiment. However the temperature increments and thus the voltage changes are very small; if the ADC is connected directly to the thermistor as in Chapter 3, the changes are less than the step size of digitization. To overcome this problem an amplifier is used to boost the voltage change. On the protoboard attached to the experimental apparatus is an amplifier using a CA3140 operational amplifier; a schematic diagram is shown in Figure 5.5. It is not necessary to understand the details of this amplifier circuit except to note that the relationship between the three voltages V_A (output) (pin 6) and V_T (pin 3) is given by

$$V_A = G(V_T - 2.38) \qquad (5.4.1)$$

For the circuit components used, the gain G is equal to 21.

The amplifier output (V_A) is constrained by the characteristics of the CA3140 to be between 0 V and +3 V. Since a rise in thermistor temperature will lead to a rise in the output voltage of the amplifier, the potentiometer R_1 should be set so that the output voltage of the circuit starts near the lowest voltage before a heat pulse is applied. This will allow the greatest voltage swing as the thermistor heats up without exceeding the 3 V limit. Using the oscilloscope to monitor the output voltage of each amplifier, set the poten-

Fig. 5.5. Voltage amplifier circuit for heat flow apparatus.

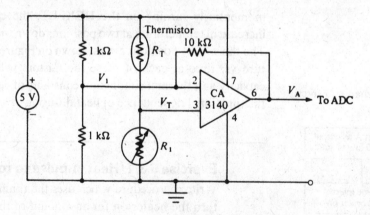

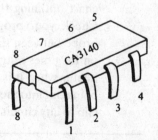

Fig. 5.6. Flow chart for Exercise 5.4.1.

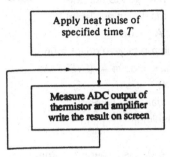

tiometers (one for each amplifier–thermistor combination) so that the amplifier outputs are about 0.20 V before you start each run. When this is done each potentiometer R_1 has been adjusted to be essentially the same resistance as the thermistor resistance R_T before a temperature pulse is applied. Since the amplifier gain is 21, the change in the output voltage ΔV_A will be 21 times greater than the change in the thermistor voltage ΔV_T.

Exercise 5.4.1 Amplifier check

Before writing a detailed program write a simple program to see that the apparatus is functioning following the outline shown in Figure 5.6. When you run this program you should see on the oscilloscope the voltage output rise and then slowly fall. It should start above 0 V and not try to go above 3 V.

Do the same to check thermistor 2, the lower thermistor. You will need to let the apparatus cool down and reset the potentiometer between heat pulses.

Exercise 5.4.2 Heat flow real-time plot

(a) The next task is to make thermistor ADC measurements at specified times. To do this, modify the program by putting in time delays as indicated by the Figure 5.7. Note that a sample is taken before the heater is turned on (A1(0) and A2(0)). This records the baseline ADC reading. The heating of the rod then changes the ADC reading from this starting value.

Fig. 5.7. Flow chart for Exercise 5.4.2.

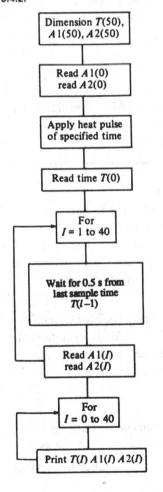

(b) Combine the data gathering procedure with a plotting procedure so that these unprocessed data are plotted as they are gathered, ie, in real-time. To do this the Turbo Pascal plotting functions in the Graph unit are used. The following program is a simple example of the graph functions. Modify it appropriately to plot your data. Scale the data properly so that they fit on the screen.

```
program GraphExample;
{shows how to use some of the graphing functions}
uses crt,graph;              {crt for Keypressed function}
                             {graph for plotting functions}
var
    GraphDriver : integer;   {these two need to be variables}
    GraphMode  : integer     {for the InitGraph procedure}
    x,y        : integer;    {screen coordinates}
    Xmax, Ymax : integer;    {Max screen coordinates in mode
                              found}
    ch         : char;       {for result of ReadKey function}
```

```
begin
  GraphDriver : = Detect;           {signals InitGraph to find the}
                                    {highest resolution available}
  InitGraph( GraphDriver, GraphMode, 'A : \Quote);
                                    {init graph mode with driver found}
                                    {on the disk in drive A: (SYSTEM)}
  Xmax : = GetMaxX; Ymax : = GetMaxY;
                                    {get boundaries of screen}
  y : = trunc(Ymax/2);              {draw baseline}
  Line(0, y, Xmax, y);
  for x : = 0 to Xmax do begin      {draw sine;0, 0 is upper left corner}
    y : = trunc(Ymax * (1 − sin(2 * Pi * x/Xmax))/2);
    PutPixel(x,y,White);
  end;
  OutTextXY(0,150,'Sine wave; Graphics mode');
  repeat until Keypressed;          {wait for key}
  ch : = ReadKey;                   {clear key from buffer}
  RestoreCRTMode;                   {go back to text display}
  writeln('now in Text mode');
  CloseGraph;
end.
```

(c) Write another procedure so that the arrays of ADC readings A1[i], A2[i], T[i] and the time the heater was on are saved on the disk as a data file when all the data has been gathered.

5.5 Data analysis

Before proceeding to more data plots and analysis, here are some additional mathematical considerations. We will assume that the temperature and voltage changes at the thermistor are small enough so that their behaviors are adequately described by differentials. Thus: (change in amplifier output voltage) = (gain) × (change in the input voltage)

$$dV_A = G dV_T \tag{5.5.1}$$

The relationship between V_T and R_T is similar to the thermistor experiment of Chapter 3, ie, $V_T/V_0 = R_1/(R_1 + R_T)$ with $V_0 = 5$ V. The relationship between dV_T and thermistor resistance changes dR_T can be obtained by differentiation; the result (which you should work out) is

$$\frac{dV_T}{V_0} = -\frac{dR_T}{R_1}\left(\frac{1}{1 + R_T/R_1}\right)^2 \tag{5.5.2}$$

Noting that R_1 and R_T are adjusted to be nearly equal at the outset gives

$$dV_T/V_0 = -dR_T/4R_T \tag{5.5.3}$$

The next task is to relate a change in the thermistor resistance to a change in temperature. The relation between thermistor resistance and temperature is $R_T = R_0 \exp(T_0/T_a)$ as discussed in Chapter 3. Differentiation of R_T with respect to temperature T_a gives

$$\frac{dR_T}{R_T} = -\frac{T_0}{T_a}\frac{dT_a}{T_a} \tag{5.5.4}$$

where T_a is the absolute temperature (K) (not the excess temperature, $T(t, z) - T_s$) and dT_a is a small temperature change due to the heat pulse. Thus if dT_a is small, it can be approximated by the measured temperature change of the apparatus (ie, the excess temperature) and T_a can be approximated by room temperature. Appropriately combining Equations (5.5.3) and (5.5.4) gives the result

$$dT_a = T_a\frac{4}{G}\frac{T_a}{T_0}\frac{dV_A}{V_0} \tag{5.5.5}$$

As Equation (5.5.5) shows, the change in output voltage in volts is not important, only its ratio with V_0. This ratio dV_A/V_0 is equal to the ratio of the change in ADC units to the ADC full scale reading.

Exercise 5.5.1 The thermal conductivity and specific heat of copper

Plot the data which you have taken with the vertical axis in temperature change from the initial temperature (Equation (5.5.5)) and the horizontal axis in seconds ($T_0 = 3440$ K for the GB32J2 thermistor). As a first step in the analysis of these data use the relations derived in Exercise 5.1.2(c), ie, visually estimate the position T_{peak} and height T_{peak} of the peak and calculate t_1 and T_1 from these values. Use these estimates to draw a curve on your graph of the data and check the fit. Then you may want to change your estimate and try another fit.

When you are satisfied with your values of T_1 and t_1 use them to calculate, via Equation (5.1.9), the diffusion constant $D = k/s$, the thermal conductivity k and the heat capacity $c = s/\rho$ where ρ is the density. Make an estimate of the error made in differential evaluation of the temperature change (Equation (5.5.5)) compared with actual temperature change. For doing this estimate, use the maximum change which can be measured using the amplifier circuit employed.

Another consideration can be applied to the data analysis. In deriving Equation (5.1.5) we assumed that the time during which the heater was on (τ) was very small in relation to the time the heat takes to diffuse down the rod (T_1), ie, it was an impulse of heat (see Appendix F). In doing your experiments this approximation is valid as long as you make $t = 0$ on your graph correspond to the midpoint of the heating time and if the heating time is less than any t_1. Appendix G gives the details.

Exercise 5.5.2 Time shift of heat flow data

Shift the time scale of your heat flow data by $\frac{1}{2}\tau$ and again estimate T_1, t_1, Ds and k. Compare with the previous values.

Often the visual fit done in Exercise 5.5.1 is a good way to analyze data especially if the model equations cannot be linearized. The solution (Equation (5.1.9)) of the heat flow equation is one such model equation. Using some tricks, however, a least squares fit can still be done by rederiving the parameter equations starting from the expression for the total error (Equation 3.7.1)):

$$e_2 = \sum_{i=1}^{N} (y_i^{\text{model}} - y_i^{\text{data}})^2 \qquad (3.7.1)$$

Exercise 5.5.3   Least squares fit to the heat flow data

(a) The model equation for heat flow in a rod (from Equation (5.1.9)):
$$T(t) = T_1(t_1/t)^{1/2} \exp(-t_1/t)$$
cannot be transformed into a linear form. Take the natural logarithm of $T(t)$ and identify the extra term that makes the result non-linear.

(b) Make a change in parameters by letting $a = T_1 t_1^{1/2}$ and $b = t_1$. Now take the natural logarithm of $T(t)$ again and by letting $A = \ln(a)$ and $Y_i = \ln(T_i)$ be the experimental measurements at points $x_i = 1/t_i$, write an expression for the error.

(c) Differentiate this error expression with respect to the new parameters A and b and set them to zero to find the minimum error. Solve the resulting equations to find A and b.

(d) What are T_1 and t_1 in terms of A and b?

(e) Write a program to use you equations to estimate the thermal conductivity and heat capacity of the rod from your data. Experiment with fitting different parts of the data since the later times may be affected by radiation or convective cooling.

5.6 Conduction, convection, and radiation

There are three mechanisms for transportation of heat: conduction, convection, and radiation. Conduction has been the subject of the chapter until now. Convection and radiation act at surfaces to remove heat from the rod thus the total heat Q in the system decreases with time.

Convective losses can be modeled as

$$\Delta Q_c = hA' \, \Delta T \, \Delta t \qquad (5.6.1)$$

where h is called the transfer coefficient (in W/m^2 K), A' is the surface area over which the convection occurs and ΔT is the difference in temperature between the surface and the air.

Radiative losses also occur when there is a difference in temperature:

$$\Delta Q_r = \sigma A'(T_{rod}^4 - T_{air}^4) \, \Delta t \qquad (5.6.2)$$

where σ is the Stefan–Boltzmann constant and A' is the area over which the radiation occurs (in this case it is the same as for the convective losses).

Exercise 5.6.1 Estimation of convective and radiative losses

(a) Show that if the temperature difference is small, the expression for radiative losses can be written

$$\Delta Q_r \approx \sigma A' 4T^3 \, \Delta T \, \Delta t \qquad (5.6.3)$$

where T is room temperature (K).

(b) Show that the combined convective and radiative losses for a cylindrical rod of radius r and length dz are

$$\Delta Q_{tot} = (h + 4\sigma T^3) \, dz \, \Delta T \, \Delta t \, 2\pi r \qquad (5.6.4)$$

Thus the combined losses for a small piece of the rod dz are proportional to dz ΔT.

(c) The total heat loss is obtained by integrating the equation of part (b) over the length of the rod.

$$Q_{tot} = \int \Delta Q_{tot} = (h + 4\sigma T^3)2\pi r \, \Delta t \int \Delta T \, dz \qquad (5.6.5)$$

That is, it is proportional to the area under the curve in Figure 5.2. Integrate Equation (5.1.8) to show that this area is equal to

$$\int_0^\infty T(z, t) \, dz = Q/As \qquad (5.6.6)$$

where A is the cross-sectional area of the rod and s is the volumetric heat capacity. Note that this result is independent of the time t so that heat is lost at a constant rate.

(d) Now show that the total convective and radiative heat loss is

$$Q_{tot} = (h + 4\sigma T^3) \frac{2Q}{rs} \, \Delta t$$

(e) Make an estimate of the heat lost for times of 2 and 10 s. The transfer coefficient h for free air is 20 W/m^2 K. What percentage of the total heat in the system is lost to these effects? Quantitatively, how do you expect these losses to affect the data and calculated parameters in the heat flow experiment?

6 IBM-PC architecture and assembly language programming

Thus far, there has been no need to understand the inner workings of the computer in order to do useful experiments. It has been a black box which responds in a reliable way when given instructions. Just as in using a car, many times this is sufficient; however, to utilize its capabilities as a tool in the laboratory fully, the internal operation of the computer should be understood. In this chapter, we will look under the hood to explore the internal organization of the IBM-PC and to learn to program the 8088 microprocessor directly.

6.1 Inside the IBM-PC

A first glance under the cover of the IBM-PC shows a circuit board with a row of connectors which contain other circuit boards standing vertically. The horizontal board (the mother board) contains the 8088 microprocessor chip and various other chips which control the keyboard and screen and contain the memory cells. The microprocessor is the CPU which controls the system and executes the program instructions. The boards in the connectors perform a variety of other functions. Figure 6.1 shows the general organization.

The different chips and circuit boards communicate with each other via the buss: a group of 62 wires which carry digital signals. There are 20 address lines, 8 data lines, and 34 auxiliary lines. The data lines contain the 8 bits of data which are to be transferred by the CPU. The bits on the address lines specify the binary number of the location from or to which the data will be transferred. The operation of the computer is, at the lowest level, a controlled transfer and manipulation of bits of data among various chips and devices.

The CPU uses the auxiliary wires to control the data transfers. The RD and WR wires signal the direction of data transfer: a read (RD) transfers data into the CPU and a write (WR) transfers data from the CPU. The IO/M wire indicates whether the address is an I/O (Port) address or regular memory (Mem) address. The CLK wire controlled by the oscillator is the clock which determines how fast the data transfers will take place. The one in the IBM-PC generates 4.77 million cycles per second (Figure 6.2). It takes four CLK cycles to transfer one byte of data (one buss cycle). During the first cycle the address is placed on the address buss along with the IO/M indicator

Fig. 6.1. Inside the IBM-PC.

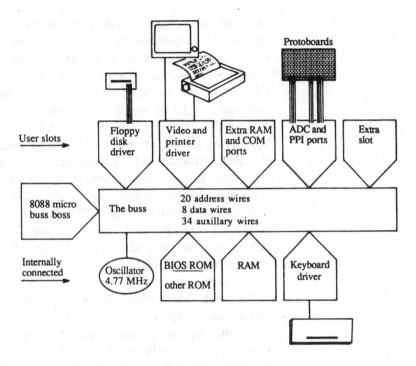

Fig. 6.2. Buss cycle timing from iAPX86, 88 user's manual. (Used by permission.) Clock cycles (e.g. T_1) occur at 4.77 MHz.

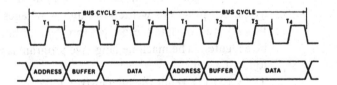

Figure 4-5. Typical BIU Bus Cycles

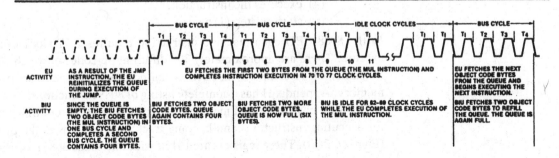

Figure 4-6. BIU Idle States

(address cycle). Since it takes the memory circuits some time to find the unique address specified, the CPU waits one cycle before proceeding (buffer cycle). During the remaining two cycles the data is transferred (data cycles). Sometimes it takes even longer for the memory or I/O port to respond to the address. In that case extra memory cycles (called wait states) are inserted between T_3 and T_4 to wait for a response.

Even though the microprocessor can access memory in four cycles (the buss cycle), the execution of many instructions can take much longer. Then the clock cycles continue while the microprocessor is busy. These are called idle clock cycles because the buss is idle, not the CPU!

As stated above, the buss has 20 address wires and 8 separate data wires. This is not true of the 8088 CPU itself. At the chip the address wires AD0–AD7 are also used as the data wires. This can be done since the address and data are used at different times in the buss cycle. There are other chips between the 8088 and the buss which separate the address and data signals onto their own wires.

The eight binary bits (or one byte) of data transferred in each buss cycle can represent various things. They could be a binary data value, machine instruction (operation code or op-code), part of a real number representation or one half of a 16-bit address. The electronic protocol for transferring the data is always the same no matter what the data may represent.

6.2 The 8088 microprocessor

Programs are ultimately stored in the computer as a series of data bytes. All programs written in other languages (eg, BASIC, FORTRAN, Pascal) are translated into this form (by another program!) before they can be executed. The machine language program is executed by the CPU by the following steps:

The CPU (1) reads the next instruction code in the series,
            (2) decodes the instruction,
            (3) if necessary, reads additional data (such as an address or data byte),
            (4) executes the instruction,
            (5) starts at step (1) again.

In the 8088 CPU each machine instruction requires 4–77 clock cycles to execute. There are approximately 300 different op-codes of which 100 are fundamental. A complete instruction can occupy from one to six bytes of memory. Appendix H has a complete listing of 8088 instructions.

The 8088 has 14 internal memory locations called registers which it uses for executing instructions and keeping track of where it is in the program (Figure 6.3(a)). These registers are 16 bits wide and so the 8088 is considered a 16-bit microprocessor by the manufacturer even though it talks to the outside world using 8 bits at a time. Machine instructions can refer to 8 or 16

Fig. 6.3. (*a*) General registers of the 8088 microprocessor. All registers are 16 bits wide but the general registers can be addressed in separate 8-bit pieces, e.g. AH, AL. (*b*) Addition of segment and offset to give a complete 20-bit address.

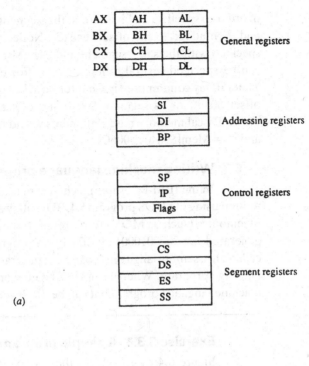

(*a*)

(*b*)

bits at a time. The CPU will automatically transfer two bytes into or out of the chip if called for.

The 14 registers are grouped by function. AX, BX, CX, and DX are general purpose registers for transferring and manipulating data. CS, DS, SS, and ES are segment registers used for determination of addresses (more on that in a moment). IP, SP, BP, SI, and DI are offset registers which are also used for addressing. In particular the IP register is used to keep track of the location in the memory of the next instruction to be executed. The remaining register, the Flag register, has individual bits (flags) which are used to indicate particular states of the CPU. For example, ZF (zero flag) indicates if the last instruction (perhaps a subtraction) had a result of zero.

Since the internal registers are 16 bits wide and addresses are 20 bits wide, it is not possible to use one register to directly address all of the memory. Two registers are used but they are not placed next to each other to form a 32-bit address. They overlap one another by 12 bits (Figure 6.3(*b*)). That is, in order to specify a 20-bit address, the segment register is shifted over 4 bits and then added to the offset register. Notice that 65 536 addresses can be specified with the same segment address. Also, the same memory location can be specified several different ways. For example, the location of the TimeOfDay counter used in Chapter 4 can be specified as segment $0000 and offset $046C or as segment $0040 and offset $006C. The corresponding Turbo Pascal memory read statements would be n := Mem[$0000 : $046C]; and n := Mem[$0040 : $006C];.

6.3 Writing machine language programs

In the IBM-PC, short machine language programs are conveniently written using the DOS program DEBUG allows you to write programs using mnemonics (such as MOV) to represent the binary machine code which is generated (eg %00010001 00110000). A program written in mnemonics is called an assembly language program; a program in binary code is a machine language program. When using DEBUG the program is actually stored as a machine language program but can be displayed in assembly language.

Exercise 6.3.1 A simple program in DEBUG

Figure 6.4(a)–(d) shows the results of the following steps; follow along in the figure as you do them yourself.

(a) If you are in Turbo Pascal exit by typing Q. At the DOS prompt type DEBUG to run the program. The prompt will now be '–'.

(b) Type 'a' and a CR to begin the assembler mode and then type in the following program:

```
mov cx,ffff
mov dx,0323
mov al,80
out dx,al
mov dx,0320
mov al,01
out dx,al
mov al,00
out dx,al
dec cx
jnz 10c
int 20
```

When at the last line, just type CR alone and program will return to the '–' prompt. These assembly instructions will be discussed in a moment. If you make a mistake, type CR to get out of assembler mode, type 'a nnn' where nnn is the offset of the instruction you want to change, then begin to enter the instructions from there again.

Fig. 6.4. Results on the screen for: (a) Exercise 6.3.1(a); (b) Exercise 6.3.1(b); (c) Exercise 6.3.1(c); (d) Exercise 6.3.1(e); (e) Exercise 6.3.2;(f) Exercise 6.3.3(a); (g) Exercise 6.3.3(b).

(a) A:\> debug

(b)
```
2C9F:0100    mov cx,ffff
2C9F:0103    mov dx,0323
2C9F:0106    mov al,80
2C9F:0108    out  dx,al
2C9F:0109    mov dx,0320
2C9F:010C    mov al,01
2C9F:010E    out  dx,al
2C9F:010F    mov al,00
2C9F:0111    out  dx,al
2C9F:0112    dec  cx
2C9F:0113    jnz  10c
2C9F:0115    int  20
2C9F:0117
```

(c) –g
Program terminated normally

(d) –u 100
```
2C9F:0100  B9FFFF     MOV    CX,FFFF
2C9F:0103  BA2303     MOV    DX,0323
2C9F:0106  B080       MOV    AL,80
2C9F:0108  EE         OUT    DX,AL
2C9F:0109  BA2003     MOV    DX,0320
2C9F:010C  B001       MOV    AL,01
2C9F:010E  EE         OUT    DX,AL
2C9F:010F  B000       MOV    AL,00
2C9F:0111  EE         OUT    DX,AL
2C9F:0112  49         DEC    CX
2C9F:0113  75F7       JNZ    010C
2C9F:0115  CD20       INT    20
2C9F:0117  20E3       AND    BL,AH
2C9F:0119  EB10       LMP    012B
2C9F:011B  90         NOP
2C9F:011C  E882E3     CALL   E4A1
2C9F:011F  E8BAE2     CALL   E3DC
```

(e) –d 100

(f) –n SqWave.bin
```
–r
AX=0000   BX=0000   CX=0000   DX=0000   SP=FFEE   BP=0000
SI=0000   DS=2C9F   ES=2C9F   SS=2C9F   CS=2C9F   IP=0100
NV UP EI PL NZ NA PO NC
2C9F:0100  B9FFFF          MOV    CX,FFFF
–r cx                      CX 0000
```

```
:20
−r
AX=0000   BX=0000   CX=0020   DX=0000   SP=FFEE   BP=0000
SI=0000   DI=0000   DS=2C9F   ES=2C9F   SS=2C9F   CS=2C9F
IP=0100   NV UP EI PL NZ NA PO NC
2C9F:0100 B9FFFF        MOV     CX,FFFF
−w 100
Writing 0020 bytes
```

(g) −n SqWave.bin
　　−l
　　−
```
u 100
2C9F:0100 B9FFFF        MOV     CX,FFFF
2C9F:0103 BA2303        MOV     DX,0323
2C9F:0106 B080          MOV     AL,80
2C9F:0108 EE            OUT     DX,AL
2C9F:0109 BA2003        MOV     DX,0320
2C9F:010C B001          MOV     AL,01
2C9F:010E EE            OUT     DX,AL
2C9F:010F B000          MOV     AL,00
2C9F:0111 EE            OUT     DX,AL
2C9F:0112 49            DEC     CX
2C9F:0113 75F7          JNZ     010C
2C9F:0115 CD20          INT     20
2C9F:0117 20E3          AND     BL,AH
2C9F:0119 EB10          LMP     012B
2C9F:011B 90            NOP
2C9F:011C E882E3        CALL    E4A1
2C9F:011F E8BAE2        CALL    E3DC
```

(c) Use the PrtSc key on your computer to get a listing of the program.
Then type 'g' CR to run the program. It should run for a second or
so, then print 'Program terminated normally' and given the prompt
'−'.

(d) Connect the oscilloscope to PA0 and try to measure the high and
low times of the waveform after typing 'g'. You will probably need
to run it a few times before getting the oscilloscope adjusted
correctly.

(e) Type 'u 100' to get a listing of the program and the machine language
bytes. Use PrtSc again to get a listing. Note the time ratio of this
program with the SquareWave you wrote before.

	Machine language	Assembly language	Turbo Pascal
Fig. 6.5. Correspondence between program statements in machine language, assembly language and Turbo Pascal.			BEGIN
	2C9F:0100 B9FFFF	MOV CX,FFFF	FOR I := 1 to $FFFF DO BEGIN
	2C9F:0103 BA2303	MOV DX,0323	PORT[803] := 128;
	2C9F:0106 B080	MOV AL,80	
	2C9F:0108 EE	OUT DX,AL	
	2C9F:0109 BA2003	MOV DX,0320	PORT[800] := 1;
	2C9F:010C B001	MOV AL,01	
	2C9F:010E EE	OUT DX,AL	
	2C9F:010F B000	MOV AL,00	PORT[800] := 0;
	2C9F:0111 EE	OUT DX,AL	
	2C9F:0112 49	DEC CX	END; {OF FOR LOOP}
	2C9F:0113 75F7	JNZ 010C	
	2C9F:0115 CD20	INT 20	END.

This program is an assembly language equivalent of the program SquareWave of Chapter 3. Figure 6.5 shows the correspondence between the statements of machine language, assembly and Turbo Pascal programs.

Find a quiet space and follow closely the next description. The assembly program begins with the instruction MOV CX,FFFF. The MOV instruction is the most used assembly language instruction since it moves data from one place to another. It has many different forms depending on where the data is coming from and going to. When this program is executed, the data $FFFF moves into the CX register of the CPU. This register is used as a loop counter. Next, $0323 is moved into the DX register. In decimal $0323 is 803. and is thus the I/O address of the control register of port A. Next, $80 (decimal 128.) is moved into the AL register (the low byte of the AX register). Then the OUT DX,AL instruction sends the data in AL to the I/O address contained in DX. The instructions OUT and IN are used to address the I/O devices. The three instructions thus have the effect of the one Turbo Pascal instruction Port[803] := 128; they initialize port A as an output port.

The next three instructions are quite similar to the last three and send $01 to the register $0320 (port A). Since the DX register already contains the correct address, only two instructions are next needed to change the port A data to $00. The next instruction DEC CX decrements the data in the CX register by 1. Thus the first time through CX will now contain $FFFFE. The next instruction JNZ 010C tests the results of the previous DEC instruction. If the result is not zero, the CPU jumps back to the instruction at the address specified, $010C. If the result is zero, the CPU continues on to the next instruction in line. Thus the CPU will loop back to pulse PA0 high and low until the DEC CX instruction succeeds in decrementing the CX register to zero; ie, the loop will be traversed $FFFF or 65 535 times. Then the INT 20 instruction returns control to the DEBUG program. In other situations this instruction would be RET (return) but if RET is used here control of the machine is eventually lost and you would need to reboot. The instruction INT 20 is one which calls the DOS function 'Program Terminate'. The remainder of the instructions which you may see on your screen are just garbage left over from the previous program.

The unassembled listing (gotten with the 'u 100' command) contains the machine language bytes as well as the assembly language text. Starting at the left, the first nine characters contain the segment offset of the instruction. The segment number could be different for your listing because it changes for different memory configurations. The offset begins at $0100 when DEBUG is started. The next set of characters are pairs of characters representing hexadecimal numbers stored in succeeding memory locations. These are the machine language instruction codes. For example, on the first line, B9 FF FF is the instruction code for MOV CX,FFFF with B9 the code for MOV CX,nnnn and FF FF the data FFFF. The second line has BA representing MOV DX,nnnn and 23 03 representing 0323. Note that the low byte of the address (23) is placed in memory first and the high byte (03) is placed after. This is a standard convention.

The machine language bytes are all that really exist in the machine when using DEBUG. The assembly language text is translated into machine language as it is entered in the 'a' command. The machine language bytes are translated into assembly language when the 'u' command is given.

Exercise 6.3.2 Memory dump

A direct dump of bytes in memory without translation can be done with the DEBUG 'd' command. Type 'd 100' and compare the direct dump with the 'u 100' results.

Using DEBUG specific bytes of memory can be saved on disk and retrieved back into memory as binary files. Thus, the program just written can be saved as a set of bytes like the ones shown by the 'd 100' command. This is different than a text file in which the Pascal programs and the data are saved. The 'type' command will produce gibberish when used on a binary file.

Exercise 6.3.3 Writing a binary file

(a) Save the program on disk as follows: give the program a name by typing 'n sqwave . bin'CR. The command 'w 100' (where 100 is the starting address) is used to save it on disk but before that is done the length of the program needs to be put in the CX register. Type 'r'CR and examine the current values in the registers especially the CX register. Now type 'r cx' and enter the number of bytes in hexa-decimal to save. For example, entering '20' will save 32 bytes which is more than is needed but OK to do. Type 'r'CR again to look at the registers. Now 'w 100' will save the program.

(b) To retrieve a program: first name it with 'n sqwave . bin' then type 'l' to load it into memory. Any part of RAM can be saved and loaded in this manner. For example, to save an image which is on the screen you could save the memory which is used to store the screen bits.

Exercise 6.3.4 Machine language square waves

Write, run, print out and save a machine language program which produces square waves on PA4. Examine the signals with the oscilloscope.

6.4 Compiled Pascal and Inline

As you have seen, even though a Pascal program is compiled into machine language before execution it runs slower than an equivalent machine language program. The compiler cannot anticipate what the program is trying to do so will insert what seem like extra machine instructions so that all possibilities will be covered. DEBUG can be used to look at a program which has been compiled.

In order to use DEBUG efficiently the memory location of the code you want to look at needs to be discovered. The Turbo Pascal function Seg(Name); will return the segment of the variable, function, or procedure Name and Ofs(Name); will return the offset.

Exercise 6.4.1 Compiled Turbo Pascal

(a) Write and run the following program:

```
program FindLocation;
{to find the location of this program in memory—
uses crt;                        {for keypressed }
procedure SqWave;
{outputs a square wave on PA0 of the 8255 as fast as it can}

const
        PortControl = 803;       {the control port for 8255 #1}
        PortA       = 800;       {the register for port A #1}
begin
    Port[PortControl] : = 128;   {set port A as output in mode 0}
                                 {ports B and C are also outputs}
```

```
    repeat
        Port[PortA] := 1;          {set PA0 high, keep PA1–PA7 low}
        Port[PortA] := 0;          {set PA0 low, keep PA1–PA7 low}
    until keypressed;
    end;

begin
    writeln('Segment ', Seg(SqWave),' Offset ', Ofs(SqWave) );
    SqWave;
end.
```

(b) To look at this program with DEBUG it must be made into a program which can be executed from DOS instead of from within the Turbo system. This is done by changing the compiler options. Type 'Alt O' to go to the options and then 'D' to change the destination from memory to a disk file. 'Alt C', 'C' will then compile the program to a disk file with the '.EXE' filename extension.

(c) Quit Turbo and start DEBUG. Retrieve the compiled program using 'n FINDLOC.EXE' and 'l'. Then use 'g' to run the program and find out where the procedure is in memory.

(d) Use the 'u nnn' command to look at the code at the offset reported by the program.

Looking at the results in Figure 6.6, it can be seen that the compiler was quite efficient in translating the Pascal code into machine code. After a few opening instructions, the code to do the actual port access is very close to that programmed in assembler in Section 6.3.

In general, however, compiled code will not be as efficient as carefully programmed assembler code. Turbo Pascal provides a means to insert machine language instructions into a Pascal program if direct control of the generated code is desired. This is done using the Inline( ); statement. The actual machine language bytes need to be used so DEBUG will be used to write the code and then the bytes are inserted into the Turbo Pascal program. There are Assembler–Turbo Pascal Inline translators available if a lot of this needs to be done.

Fig. 6.6. Results of Exercise 6.4.1.

	Results		Comments
	—n findloc.exe		
	—l		
	—g		
	Segment 11798 Offset 0		this is $2E16
	Program terminated normally		
	—u 2e16:0000		
2E16:0000	55	PUSH BP	begin SqWave procedure
2E16:0001	89E5	MOV BP,SP	preliminary step.
2E16:0003	31C0	XOR AX,AX	preliminary step
2E16:0005	9AAD02862E	CALL 2E86:02AD	preliminary step
2E16:000A	B080	MOV AL,80	this is 128 decimal
2E16:000C	BA2303	MOV DX,0323	this is 803 decimal
2E16:000F	EE	OUT DX,AL	
2E16:0010	B001	MOV AL,01	
2E16:0012	BA2003	MOV DX,0320	
2E16:0015	EE	OUT DX,AL	
2E16:0016	B000	MOV AL,00	
2E16:0018	BA2003	MOV DX,0320	
2E16:001B	EE	OUT DX,AL	
2E16:001C	9A4503202E	CALL 2E20:0345	this must be the address for Keypressed function
2E16:0021	08C0	OR AL,AL	
2E16:0023	74EB	JZ 0010	jump back if not Keypressed
2E16:0025	89EC	MOV SP,BP	clean up
2E16:0027	5D	POP BP	clean up
2E16:0028	C3	RET	end SqWave procedure
2E16:0029	085365	OR [BP+DI+65],DL	

Exercise 6.4.2 Example Inline program

As an example of how Inline( ); works, enter and run the following program:

```
program ADCInline;
{Show use of Inline statements in program ADCTest}
{does continuous conversions of the voltage on channel 2}
{of the ADC on the John Bell Engineering Universal I/O board at base}
{address 800 ($320)}
uses crt;                   {this statement is necessary because}
                            {the crt unit contains the procedure}
                            {delay(n : integer); used below}


const
        ADCRegister = 812;
                            {the I/O register of the ADC}
        ChannelNo   = 2;
                            {the channel for conversion}


var
        ADCUnits : byte;    {conversion results in ADC units}
                            {0 to 255}


begin
  Repeat
        Inline              {this Inline is equivalent to Port[812] : = 2;}
        $BA/$2C/$03/        {mov DX,$032C}
        $B0$02/             {mov AL,02}
        $EE);               {out DX,AL}

        Delay(2);           {Wait for ADC}

        Inline(             {equivalent to ADCUnits : = Port[812];}
        $BA/$2C/$03/        {mov DX,$032C}
        $EC/                {in AL,DX}
        $A2/ADCUnits);      {mov ADCUnits,AL}

        writeln( ADCUnits : 8 );
                            {display results on screen}
        Delay (500);        {wait approx ½ s before doing the next}
                            {so the screen doesn't scroll too fast}
     until Keypressed;      {loop continuously until a key is pressed}
end.
```

Exercise 6.4.3 Inline square wave program

Using the machine language bytes from the program you wrote using DEBUG to output square waves on PA4 (Exercise 6.3.4), write a Turbo Pascal program with Inline( ); steps to output on PA4. Measure the speed of execution using the oscilloscope and compare with previous determinations.

6.5 Operation of a DAC

Let's take a short respite from learning about the inner workings of the computer and explore the use of a Digital to Analog Converter (DAC).

You have used an ADC in the previous sections to convert an analog voltage signal external to the computer into a digital signal which the computer can manipulate and store. The inverse operation is done with a DAC. The DAC is an output device which converts the binary number to an analog voltage. They can be used for a variety of purposes. For example, they are used as the output devices for digital music playback and for digital video players. You will use them to drive oscilloscope displays.

There are two DACs connected to the parallel interface, Figure 6.7; one is on Port A and one is on Port B of the second 8255 PPI. They are used by setting up the ports as output and then writing digital numbers into the ports.

For electronic reasons which need not concern us, the DAC you are using uses an inverted representation of numbers, ie, the binary number $00 at its input produces +5 V at its output and $FF at its input +0 V at the output. To generate the conventional conversion between analog voltage and binary numbers, the binary numbers at the output to the DAC should be inverted, ie, all ones converted to zeroes and zeroes to ones.

Fig. 6.7. DAC circuit.

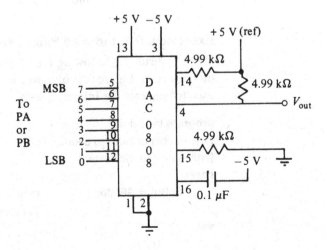

Exercise 6.5.1 DAC sawtooth wave

Write a Pascal program which will set up port A for output and write the temporal sequence of numbers $00, $01, ..., $FF, $00, ... *ad infinitum* into the port. Observe the output of the DAC (pin 4) with the oscilloscope and note the results.

Exercise 6.5.2 DAC sine wave

Write a program which will output a sine wave from the DAC. Take note of the fact that the sine function goes from −1 to +1; this must be put in a digital range from 0 to 255 for the DAC. Display the sine function in two ways: (*a*) by calculating the sine each time it is needed, (*b*) by using a lookup table. In (*b*) a table (array) of 100 sine values is calculated once and then, when the program needs a value, it is obtained from the array. Observe with the oscilloscope and note the difference in speed of the two methods of programming. To convert the real number calculations to integers or bytes which are needed for the Port[ ] statement, use the functions Trunc( ); or Round( );.

6.6 Indexed addressing

In order to output table of values (like the table in Exercise 6.5.1) using machine language, indexed addressing is used. Indexed addressing in machine language is the equivalent of using array elements in a Pascal program. A simple Pascal program to move successive array elements to the display screen buffer is shown in Exercise 6.6.1. The statement transfers 100 array elements to the screen buffer memory locations 1000, 1001, 1002, ..., 1099. This will produce an odd array of symbols on the screen just to show that it is indeed working. Exercise 6.6.1 also shows a DEBUG listing and a Turbo Pascal Inline program which will do the same thing.

Exercise 6.6.1 Indexed Pascal and DEBUG programs

Write and run the following Pascal programs. Observe the output on the screen. Figure 6.8 contains a listing of the DEBUG program used to generate the Inline( ); data.

```
program IndexPas;
{an example of indexed addressing}
{writes junk to the screen}
const
     ScBuf = $B800;              {this is for the color screen}
                                 {for monochrome screen use $B000}
var
     i : integer;
```

```
begin
    writeln('Program IndexPas'); {scroll screen}
    for i:= 0 to 100 do
        Mem[ScBuf:1000+i] := i;   {write to middle of screen buffer}
end.
```

```
program IndexInline;
{program to show indexed addressing}
{writes monkey business to screen}
var
    NData : integer;            {number of data elements in the array}
    FunnyData : array[0..1000] of byte;
                                {array data}
    i : integer;                {a counter}
begin
    writeln('Program IndexInline');   {scroll screen}
    NData := 100;
    For i:= 1 to NData do
        FunnyData[i] := i;
    Inline(
        $8B/$1E/NData/          {MOV BX,[NData]   inits counter in BX}
        $B8/$40/$B8/            {MOV AX,$B840    $B800 is start of}
                                {color screen}
                                {use $B040 for monochrome screen}
        $8E/$C0/                {MOV ES,AX   put into ES register}
        $8A/$87/FunnyData/      {* MOV AL,[BX+FunnyData]   get data}
                                {from array}
        $26/$88/$07/            {ES : MOV [BX],AL   write data to}
                                {screen buffer}
        $4B/                    {DEC BX   decrement counter}
        $75/$F6 );              {JNZ *   if not done}
end.
```

Fig. 6.8. DEBUG program for
Exercise 6.6.1.

```
-u 100
60E4:0100 8B1EFE7F     MOV    BX,[7FFE]
60E4:0104 B840B8       MOV    AX,B840
60E4:0107 8EC0         MOV    ES,AX
60E4:0109 8A87FE7F     MOV    AL,[BX+7FFE]
60E4:010D 26           ES:
60E4:010E 8807         MOV    [BX],AL
60E4:0110 4B           DEC    BX
60E4:0111 75F6         JNZ    0109
60E4:0113 CD20         INT    20
```

These programs illustrate several new features of machine language programming and Inline( ); statements. The first DEBUG statement (Figure 6.8) is a new way to use MOV. It instructs the CPU to take a two-byte number starting at the memory location $7FFE and put it in the BX register. In most cases when memory is addressed the DS (data segment) register is used for the segment part of the address. If the instruction were executed, the two bytes at memory DS : $7FFE and DS : $7FFF would be placed in the low and high bytes of the BX register. The brackets are used to indicate that the contents of the memory locations $7FFE–$7FFF should go into the register not the number $7FFE itself. The CPU knows to transfer two bytes because the register referred to (the BX register) is a two-byte register.

In reality, the number $7FFE is used as a place holder so that DEBUG will generate the correct code for a two-byte offset. In the Inline listing the $7FFE has been replaced by a variable, NData. The Turbo compiler will place the two-byte offset of the variable NData in the code at that spot when the final code is generated. So, the MOV instruction will get the value of the variable NData and place it in the BX register. This provides a nice means of referring to Pascal variables from within Inline. Simple variables are stored by Turbo in the data segment so the correct segment is automatically used in this case. In this program the BX register is used as a loop counter.

The next two instructions are used together to place the segment address of the screen buffer in the ES (extra segment) register. There is no direct access to the ES register from memory so the number is first put in the AX register then transferred to the ES register. Notice there are no brackets around the number $B840 since it is that number (not the number in that memory location) which should go into the AX and ES registers.

The information displayed on the video screen is kept in RAM. It can begin either at segment $B000 or $B800 depending on the computer used. The bytes in this memory area are scanned to produce the image on the screen. If any of the memory is changed then the screen image is changed as seen when the programs are run.

The next MOV is again a new form of the instruction. It says load the AL register with the byte found at the address calculated by adding $7FFE to the contents of the BX register. As before the DS segment is assumed. The BX register is a counter (as will be evident in a moment) so this instruction accesses different memory locations as BX is changed, ie, it can access an array of data if $7FFE is the beginning of the array. Again the number $7FFE is used as a place holder in the DEBUG program so that a variable can be used in the Inline program. In the Inline program the name of the array is placed where the two bytes would be and the compiler places the address of the beginning of the array at that point in the code.

The next MOV is yet another form of the instruction. It says to put the byte from the AL register into the address given by the BX register. However, the instruction is preceded by the ES: mnemonic which says use the ES register for the segment portion of the address instead of the usual DS register. This is called a segment override. The brackets around BX indicate that the byte should be placed at the address contained in the register not into the register itself.

Finally the next two instructions are more recognizable. The BX register is decremented (DEC BX) and tested to see if it is zero (JNZ); if it isn't then the CPU jumps back to offset 109 in DEBUG and * in the Inline listing.

The net result is that the number NData is placed in the BX register, then successive bytes of the array FunnyData are put in the AL register and written to successive memory locations in the screen buffer. Each time through the loop the BX register is decremented and tested until it reaches zero and the program stops.

Exercise 6.6.2 DAC output Inline

Write the data output portion of the DAC program of Exercise 6.5.2(b) (lookup table version) using Inline statements with indexed addressing. Use a Repeat . . . until Keypressed; construct around the Inline portion so that the program can be stopped without rebooting.

6.7 An X–Y plotter

By using two DACs and the oscilloscope you can make an X–Y plotter, ie a display whose X value is determined by one function and whose Y value by another function of the same parameter. The oscilloscope will display the two input channels in this way if you set the MODE to 'X–Y'. As an example of X–Y plotting, suppose the x axis voltage varied as $\cos(\theta)$ and the y axis voltage varies as $\sin(\theta)$, what would be the figure traced out as successive points were plotted ($\theta = \theta_1, \theta_2, \ldots$)?

Exercise 6.7.1 Lissajous figures on a DAC X–Y plotter

Use the two ports and DACs to plot Lissajous figures.

(a) Begin with the simplest figures, a circle:
$$x = \cos(\theta), \ y = \sin(\theta)$$
and a line:
$$x = \cos(\theta), \ y = \cos(\theta)$$

(b) Next try:
$$x = \cos(\theta_1), \ y = \sin(\theta_2)$$
where
$$\theta_1 = 2\theta_2 \quad \text{or} \quad \theta_2 = 2\theta_1$$

(c) What happens when you vary the relative phase or amplitude of x and y? For example, try a circle again but with

$$x = \cos(\theta), \quad y = \tfrac{1}{2}\sin(\theta)$$

then

$$x = \cos(\theta + \tfrac{1}{4}\pi), \quad y = \tfrac{1}{2}\sin(\theta)$$

6.8 Boolean algebra

Normal algebraic variables can take on an infinity of values and are added, subtracted, multiplied, etc. to give new values. Boolean variables are quantities which can take on only two values and are operated upon by AND, OR, NOT, etc to give new values. The two values can be described by 0 and 1, high and low, or true and false. (No $\tfrac{1}{2}$, middle, or maybe.) The AND operation combines two Boolean variables A and B to produce a third Boolean variable C such that C is 1 if, and only if, both A and B are 1. The AND operation between two Boolean variables is represented by $\wedge$ or by a dot,

$$C = A \cdot B \quad or \quad C = A \wedge B$$

Boolean algebra statements are frequently defined by truth tables. Table 6.1 shows the AND operation

Table 6.1 *Truth table for the* AND *operation*

A	B	$C = A \cdot B$
0	0	0
0	1	0
1	0	0
1	1	1

The 8088 has an instruction AND which does exactly this. Each of the data bits is considered as a separate Boolean variable. The AND instruction performs the AND operation between each of the corresponding bits in the named register and a memory location and deposits the result in the register. Turbo Pascal also has an And operator. If it is used between Integer or Byte variables it will act like the machine language instruction, ie, it will AND each pair of bits independently. It can also be used between variables declared Boolean and will then produce only one Boolean result.

An important application of the AND instruction is to help determine whether some particular bit in a byte or word (byte pair) is a 0 or 1. The first step is to isolate the bit being tested and then to produce a result dependent only on the value of the bit. This operation is called masking. It is as though we hide the bits of no concern behind a mask and look through a hole in it at the one of interest. The result is independent of the value of the other bits.

For example, the program steps

```
MOV BX,0200
MOV AL,[BX]
AND AL,10
MOV [BX],AL
```

will isolate bit 4 of the contents of the memory location at offset $0200 and put the result back in $0200. The Turbo Pascal statement which does the same thing is

Port[$0200] : = Port[$0200] And $10;

The AND operation between the number loaded into the AL register and the 1 in bit 4 of the number $10 will produce a 1 in bit 4 of the result if bit 4 of AL was 1 and will produce a 0 if bit 4 was a 0. All of the data bits of the result will be 0 because 0 AND 0 or 0 AND 1 is always 0.

Exercise 6.8.1 AND

(a) To see how this works, use DEBUG to write the machine language code for the following code

```
MOV BX,0200
MOV AL,[BX]
AND AL,[BX+1]
MOV [BX+2],AL
INT 20
```

and use the DEBUG E (enter) command to store some hexadecimal numbers at offsets $0200 and $0201. Run the program (g = 100) and examine the results with the D (dump) command. Verify by hand, by translating to binary notation, that the AND operation was done correctly. Use starting numbers so that the entire AND truth table is verified.

(b) The AND operation can also be used to set particular bits to 0. Try the program with $FE in $0200.

The Boolean algebra operation conjugate to AND is OR, which given two Boolean variables A and B, will produce a Boolean variable C which is 1 if A or B is 1. The OR operation is written $C = A + B$ or $C = A \vee B$. It is defined by the truth table, Table 6.2.

Table 6.2 *Truth table for the* OR *operation*

A	B	$C = A + B$
0	0	0
0	1	1
1	0	1
1	1	1

Exercise 6.8.2 OR

(a) Rewrite the program in Exercise 6.8.1 using the OR instruction in place of the AND. Run the program and write out in binary form the resulting byte in $0202. Use starting numbers which verify the entire table.

(b) The OR operation can also be used to set particular bits to one. Try the program with $01 in $0200.

The third Boolean algebra instruction is the Exclusive OR (XOR). It is used just like the AND and OR operations. XOR is written

$$C = A \oplus B \qquad \text{or} \qquad C = A \veebar B$$

The rule for XOR between A and B is that C will be high only if either A is high or B is high not both. Table 6.3 is the truth table for XOR.

Table 6.3 *Truth table for the* XOR *operation*

A	B	$C = A \oplus B$
0	0	0
0	1	1
1	0	1
1	1	0

An important application of XOR is to invert one or more bits in a memory location and leave all the others alone. For example, you could produce square waves on PA7 at the same time that the other lines on this port are used for other applications. This inverting property of XOR, that $1 \oplus DB = \overline{DB}$ and $0 \oplus DB = DB$; is easily derived by inspection of the truth table above. ($\overline{DB}$ means DB inverted or 'NOT DB' thus, if DB = 1 then $\overline{DB}$ = 0 and if DB = 0 then $\overline{DB}$ = 1.)

Exercise 6.8.3 XOR

Write a program with DEBUG which inverts bit 6 of the data stored in the offset $0201 and leaves the rest of the bits alone. Run it and demonstrate this property by displaying and printing out the contents of offset $0201 before and after running with several initial values.

There is another Boolean operation which is useful. It is the NOT instruction. It is different from previous operations in that it acts only on a single Boolean variable not between two Boolean variables. As shown above the NOT operation is sometimes written as a bar over the variable. NOT reverses the state of the variable on which it acts. When acting on a byte or word, NOT will invert all the bits; thus, NOT $10101100 is $01010011.

Exercise 6.8.4 NOT

The assembler instruction NOT AL will invert the bits in the AL register. The instruction NOT 0200 will invert the data at offset $0200. Write a program using DEBUG which inverts the data at offset $0200. Load bytes and examine the results.

Exercise 6.8.5 Turbo Pascal Boolean expressions

Using Turbo Pascal Boolean operations And, Or, Xor and Not on Integer or Byte variables, verify the truth tables. Note that there is no easy way to print numbers in hexadecimal using Turbo formatting commands so you will need to translate decimal numbers to their hexadecimal or binary equivalents or output them to the LED display.

6.9 Branching instructions

The first machine language program you wrote (Exercise 6.3.1) had an instruction JNZ which jumped back to the middle of the program if the result of the previous instruction (DEC CX) was not zero (Jump if Not Zero). Otherwise the CPU proceeds with the next instruction in line. This is like the Pascal construct

Repeat . . . Until (x = 0);

which will jump back to the start of the code as long as x is not zero.

The 8088 has a whole series of instructions which will cause the program to branch from the sequential path if certain conditions test true. There is also an unconditional jump (JMP) which will always branch. The Flag register is used to store the conditions which are tested. For example, bit 6 in the Flag register (ZF) is 1 if the last instruction which affects the flag had a zero result. If the result was not zero, then ZF will be 0. The JNZ instruction then tests the ZF flag and branches if it is 0. There is also a JZ (Jump if Zero) instruction which branches if ZF is 1.

Not all instructions change the condition of the ZF flag. For example, DEC and INC do but MOV and JNZ do not. Thus the conditional information can be carried over several instructions before it is tested. There are several other flag bits and many other branching instructions which can be used. The description of the assembly language instructions in Appendix H indicates which flags are affected by which instructions.

One common use of logical and branching instructions is for testing digital

Fig. 6.9. Push button circuit.

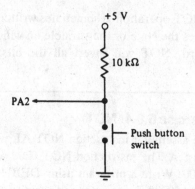

inputs. To show how this works, the circuit of Figure 6.9 will be used. Before wiring it to port C it is important to make sure that the port is set as an input. A quick way to do that is to reset the computer (reboot). That sets all the 8255 ports as inputs automatically as a precaution. You could also write and run a short Pascal program which sets all the ports as inputs. If the push button circuit were ever connected to a port which is an output, then both the push button and the chip would be trying to establish the voltage on the wire. The result would be that the 8255 chip would overheat and burn out.

Exercise 6.9.1 Masking and branching

After making sure that the port is an input, wire the circuit shown in Figure 6.9 to port C bit 0.

(a) Use DEBUG to enter and run the following program

```
        MOV DX,0322
*       IN AL,DX
        AND AL,01
        JNZ *
        INT 20
```

Here the symbol * is used to indicate where the jump should go; you will have to use a real number after the JNZ when writing in DEBUG. The program should wait in the loop until you press the button, then break out of the loop and return to DEBUG. When done use the 'r' command to display the registers. Examine them to see that they correspond to what you expect. Be sure to get a listing of the program by using the 'u' command.

(b) Write the same program using Turbo Pascal but not using Inline( ); statements; use Turbo Boolean operators.

The assembler program of Exercise 6.9.1 is simple. The data from port C by the IN instruction indicates whether the wires have 0 or +5 V on them. If the push button is not pressed the circuit keeps wire of bit 0 at +5 V and thus bit 0 will be 1. The other port C wires are not controlled by any circuits so they could be in any state. (They will probably float high.) The AND instruction with the mask $01 then isolates bit 0 and the jump (JNZ) is taken since the result of the AND is $01. Thus the program will continuously loop. When the push button is pushed, the voltage on the wire is 0 V, bit 0 is zero, the result of the AND is $00, the branch is not taken, and the INT 20 finishes the program.

An important point to note is that relative addressing is used by the 8088 in executing branch instructions. In typing the branch instruction for the push button program into DEBUG, you typed JNZ followed by the memory location of the instruction where the branch should go if true. Look at the listing of the unassembled listing of the code generated for that instruction. Note that the number $75, which is the machine code for the instruction JNZ, is first. The following byte is $FB not the number $103 which is the place where the jump is to go. The number $FB actually stands for −5 and the instruction actually means jump 5 bytes back. The CPU does this by taking the number and adding it to the current Instruction Pointer (IP) and resumes execution at that point. The IP always points to the next instruction to be executed so that in this program the IP will be at $0108 when the JNZ instruction is executed. A jump of −5 will put it back to $0103.

This scheme of relative addressing has the important consequence that program codes which have branches can be run anywhere in memory, ie, they are relocatable. Since the largest negative byte is $80 (−128) and the largest positive is $7F (127), they have the disadvantages that branches can only be taken which are no more than 128 earlier in the program or 127 bytes later. The unconditional jump (JMP) allows larger jumps and so a combination can be used if necessary. In practice this is rarely needed.

6.10 Subroutines and use of the stack

Another program branching capability which every computer must have is that of executing subroutines (procedures). A subroutine is a sequence of program steps that can be used anywhere in a program by a jump to subroutine (CALL) instruction.

To execute a subroutine, the computer stops fetching instructions sequentially from memory, jumps to the memory indicated by the call instruction and from there continues fetching instructions until a return from subroutine (RET) instruction is encountered. It then returns to the original program and resumes fetching instructions in sequential progression where it left off when the subroutine was called. This process is illustrated in Figure 6.10.

In order for the computer to return to the correct place in the calling program, the memory location of the next instruction after the CALL in the calling program needs to be saved. When the CALL NEAR instruction is executed, the 8088 stores the current IP on the stack. This is the memory

Fig. 6.10. An example of a subroutine execution sequence.

location of the next instruction after the CALL. This whole operation is analogous to writing the return address on a card and placing it on top of a pile. The last instruction of every subroutine is RET which means return from subroutine. This instruction effectively takes the top card from the pile, reads the return address, puts that location into the instruction pointer IP and then throws the card away. A CALL FAR is used to go outside a segment and both the segment and offset of the return location is saved on the stack.

The idea of using a stack (the pile of reminder cards) to store addresses may seem like a tortuous way of doing things. It is, however, an invention which was very important for the development of modern computers. Without something like a stack it is not possible to use ROM to store subroutines. One of the most awkward things about using the first successful minicomputer (the DEC PDP8) was that it did not have a stack. The stack is akin to the Reverse Polish notation used by Hewlett Packard calculators. The last item stored in the stack is the first to be retrieved. In addition to storing the return address for the subroutine, the stack is sometimes used (with care and understanding) to pass numbers from a calling program to a subroutine and vice versa. Pascal procedures and functions are subroutines and pass variables and constants on the stack. The details of the protocol are described in the Turbo Pascal manual.

The location of the stack is kept in the SS (the segment) and SP (the offset) registers. The bookkeeping of the stack is generally automatic in the CPU. From the programming point of view it is necessary that there is a RET NEAR for each CALL NEAR and a RET FAR for each CALL FAR.

Exercise 6.10.1 CALL

Write a program using DEBUG which makes two CALLS to a subroutine which waits for you to push a switch which is wired to port C2.

6.11 ADC EOC

Until now a Delay( ); statement has been inserted after starting the ADC so that it has time to do the conversion and show valid data. The ADC also has a wire to signal when the data is valid. The EOC (End Of Conversion) wire is connected to bit 7 at address $320 + $10 (see Appendix D). When it is low it indicates the converter is busy converting the data and when it is high it indicates that the data is valid. So to increase the rate at which conversions can be done, it is more efficient to test the condition of the EOC and get the data when it is ready instead of waiting some constant length of time.

Exercise 6.11.1 ADC maximum sample rate

(*a*) Write a procedure with the channel number as input and the ADC data as output which uses the EOC signal to wait for the ADC to complete the conversion. Note: the John Bell Engineering ADC is a bit slow in responding to the start signal so two wait loops are required; one to wait for the EOC to go low (indicating a Busy ADC) and another to wait for it to go high (End Of Conversion).

(*b*) Write the wait instructions of part (*a*) using an Inline( ); statement.

(*c*) See Exercise 7.3.5 for a continuation of this exercise.

7 Viscosity measurement

A solid body moving through a fluid has a force pushing on it which depends on the type of fluid. You might imagine that it would be much harder to swim in honey than it is in water. The parameter which describes this difference is the viscosity (μ). The drag force also depends upon other parameters such as the surface area of the body and the fluid density, as you will discover in this chapter. The computer will be programmed to measure the speed of a sphere falling through glycerine and to calculate the viscosity. The measurements are made with photosensors and using machine language programming.

7.1 Force required to move a solid body through a fluid

In this section the physics of a sphere moving in a fluid will be discussed. There are two distinct regimes; if the sphere is moving slowly, the dominant force resisting its motion is due to viscosity. For rapid movement, the inertial resistance of the fluid due to its density is the dominant factor. The magnitude of the resistance and the functional dependence on sphere size, velocity, fluid density and viscosity can be estimated in a rough way for both cases. This gives insight into how the drag force behaves without getting lost in the mathematics. Indeed, with turbulent phenomena exact computations have not been possible.

Viscous resistance of a fluid arises from shear in the velocity profile of flow. If two flat plates have fluid between them, as shown in Figure 7.1, a force is required to move the top one at a constant speed in relation to the bottom one. The force is proportional to the area of the plate and (if the fluid is characterized by a Newtonian viscosity coefficient) to the relative velocity and inverse distance between plates, ie, to the velocity gradient dv_z/dx.

Fig. 7.1. Drag force of a fluid on thin plates, $F_z = -A\mu(dv_z/dx)$. For a 'Newtonian' fluid, the shear force per unit area is proportional to the shear in the velocity, dv_z/dx. The viscosity μ is the proportionality constant.

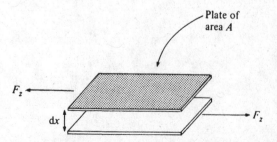

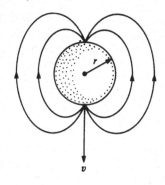

Fig. 7.2. A sphere falling slowly in a fluid, the fluid flow to move fluid from front of the sphere to the rear extends to about r away from sphere. So, $dv_z/dx \approx v/r$ and $F \approx 4\pi\mu rv$ with $4\pi r^2$ as the area of sphere.

Without doing elaborate computations this simple concept can be used to estimate the viscous resistance of a falling sphere. the effective area of velocity shear is more or less the area of the sphere, $4\pi r^2$ (Figure 7.2). The velocity perturbation resulting from moving the ball through the fluid extends to a distance about equal to the radius of the sphere; thus, the velocity gradient, dv_z/dx, which enters into the viscous drag relation is approximately v/r. Putting these two rough estimates together, an estimate of the viscous drag F_v on the sphere is

$$F_v \approx 4\pi r^2 \mu v/r = 4\pi\mu rv \qquad (7.1.1)$$

where μ is the viscosity of the fluid.

This problem is amenable to exact mathematical analysis; it was first done by Stokes and the relation is known as Stokes' law for the viscous resistance of a sphere moving in a fluid. His result† is

$$F_{\text{Stokes}} = 6\pi\mu rv \qquad (7.1.2)$$

Stokes' law is verified experimentally for cases when the sphere's motion is sufficiently slow. The approximate approach used above gives important insight into the physical origin of the Stokes' formula.

More rapid motion leads to a turbulent wake behind the sphere. Though mathematical computation of the drag force in this regime has not been done, relatively simple ideas give a good estimate of the force observed. To move an object rapidly, the speed of the fluid in the path of motion is accelerated from zero to the speed of the sphere and the fluid is pushed aside and then forms a turbulent wake behind the sphere. The turbulence eventually dissipates the kinetic energy of the moving fluid as heat and sound energy without giving any kinetic energy back to the sphere. The drag force on the sphere will be equal to the force required to push the fluid out of the way.

An estimate of the mass of fluid moved per unit time is the mass of the column of pushed aside fluid each second as the sphere falls. This is the product of the cross-sectional area A of the object perpendicular to the direction of motion, the velocity of motion v, and the density ρ of the fluid (Figure 7.3). A maximum guess is that each element of this column is accelerated to the velocity of the moving object by the pressure exerted on the front face of the object.

Therefore the work done by the drag force on the sphere (force × distance) is equal to the kinetic energy of the fluid ($\frac{1}{2}$ × mass of fluid moved × v^2).

$$(F_{\text{est}})(v\Delta t) = \tfrac{1}{2}(\rho_f \pi r^2 v\Delta t)v^2$$

Thus

$$F_{\text{est}} = \tfrac{1}{2}\rho_f v^2 A \qquad (7.1.3)$$

is the *estimated* drag on the sphere where A is the cross-sectional area.

† See, for instance, *Geodynamics: Applications of Continuum Mechanics to Geological Problems*, D. L. Turcotte & G. Schubert, Wiley, New York, 1982.

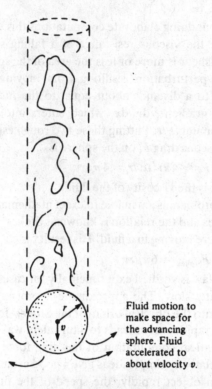

Fig. 7.3. A sphere falling with velocity v and a turbulent wake. The fluid is accelerated to about velocity v. The volume of fluid displaced each second is $\pi r^2 v$, the cross-sectional area A is πr^2.

Fluid motion to make space for the advancing sphere. Fluid accelerated to about velocity v.

The drag resistance of a blunt object in terms of an *experimentally* determined drag coefficient C_d is by definition

$$F_{drag} = C_D A \rho v^2/2 \tag{7.1.4}$$

The combination $\rho v^2/2$ is called the kinetic pressure of a fluid. The experimentally determined drag coefficient for a sphere moving rapidly through a fluid is $C_d = 0.5$. As you can see, Equation (7.1.3) overestimates the drag on a sphere by a factor of 2. Drag coefficients for other shapes are given in Figure 7.4.

Combining the Stokes relation with the turbulent force gives the total drag force on the falling object as

$$F_{tot} = 6\pi\mu r v + C_D \pi r^2 \rho v^2/2 \tag{7.1.5}$$

As Equation (7.1.4) shows, the turbulent drag for a sphere is proportional to the square of the velocity; therefore, it is the dominant phenomenon at high velocity whereas viscous drag is more important for a slowly moving sphere.

The ratio of the turbulent drag force for a sphere to the viscous drag is

$$\frac{F_{turb}}{F_{vis}} = C_D \pi r^2 \frac{\rho v^2}{2} \frac{1}{6\pi\mu r v} = \frac{C_D}{24} \frac{\rho(2r)v}{\mu} = \frac{C_D}{24} Re \tag{7.1.6}$$

$$Re = \frac{\rho 2 r v}{\mu} \tag{7.1.7}$$

The parameter Re (dimensionless) is called the Reynolds number; it is used

Fig. 7.4. Experimental drag coefficients (from p. 11.68 of *Mark's Standard Handbook for Mechanical Engineers*, ed. T. Beaumeister, 8th edn, McGraw-Hill, New York, 1978 – used by permission).

Drag Coefficients

Object	Proportions	Attitude	C_D
Rectangular plate, sides a and b	$\dfrac{a}{b} =$ 1, 4, 8, 12.5, 25, 50, ∞		1.16, 1.17, 1.23, 1.34, 1.57, 1.76, 2.00
Two disks, spaced a distance l apart	$\dfrac{l}{d} =$ 1, 1.5, 2, 3		0.93, 0.78, 1.04, 1.52
Cylinder	$\dfrac{l}{d} =$ 1, 2, 4, 7		0.91, 0.85, 0.87, 0.99
Circular disk			1.11
Hemispherical cup, open back			0.41
Hemispherical cup, open front, parachute			1.35
Cone, closed base			$\alpha = 60°, 0.51$ $\alpha = 30°, 0.34$

as a measure of the turbulence of the fluid flow. The length $(2r)$ used in defining Re for a given body is usually taken as the length of the chord in the direction of motion. Thus, for a sphere it is the diameter.

Setting Equation (7.1.6) equal to 1, shows that the change from smooth to turbulent flow occurs at a Reynolds number of about 48 (with $C_D = 0.5$). Figure 7.5 is a graph of the drag force vs Reynolds number for the range of

Fig. 7.5. Drag force vs. Reynolds number (from Turcotte & Schubert, *Geodynamics: Application of Continuum Physics to Geological Problems*, Wiley & Sons, New York, 1982).

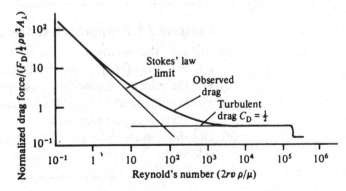

Reynolds numbers from 10^{-1} to 10^6 and shows that the transition occurs over a wide range of Reynolds numbers. The smooth flow regime is generally below a Reynolds number of 1 and the turbulent regime above 10^3.

Exercise 7.1.1 Stokes' law

(a) For a 2 mm diameter bubble of air rising through glycerine, what is the predicted terminal velocity assuming Stokes' flow? Is this what you observe in the laboratory? What is the Reynolds number? Does it agree with the assumption of Stokes' flow?

(b) By using a propeller-like flagella an *E. coli* bacterium 1 μm in diameter can swim about 0.03 mm/s in water. What is the Reynolds number? What is the drag force on the bacterium? If the bacterium can obtain 3×10^{-12} erg per molecule of glucose and can use 10% of that energy for propulsion, how many molecules per second must it metabolize to swim continuously?

Material	Viscosity (kg/m s)	Density (g/cm³)
glycerine	2.33 (at. 288 K)	1.24
air	1.78×10^{-5}	1.23×10^{-3}
water	1.0×10^{-3}	1.0

A sphere starting from rest in a liquid will be acted upon by gravity F_g and buoyancy F_b forces. Once it begins to move, the drag force Fd will act to slow its acceleration. By Newton's laws

$$F_g - F_b - F_f = ma \qquad (7.1.8)$$

F_g and F_b are constant regardless of the speed of the ball but F_d is dependent on the speed. If Stokes' flow is assumed, Equation (7.1.7) becomes a differential equation for the velocity of the sphere

$$m(dv/dt) + (6\pi r\mu)v - (F_g - F_b) = 0 \qquad (7.1.9)$$

Exercise 7.1.2 Approach to terminal velocity

(a) Assume the solution to Equation (7.1.9) is of the form $v = a[1 - \exp(t/b)]$. Plug into Equation (7.1.9) and find a and b.

(b) Plot the velocity vs. time for a glass sphere of diameter 0.60 cm and weight 0.26 g starting from rest in glycerine. What is the decay time b of the accelerated motion?

(c) How far will the sphere fall before attaining 95% of the terminal velocity?

7.2 The experimental apparatus

To measure the viscosity of a fluid the apparatus like that shown in Figure 7.6 will be used. It consists of a column of glycerine into which spheres of various sizes and compositions can be dropped and observed to fall under the influence of gravity. The velocity of the falling sphere can be measured by noting the time at which it moves through each of the four light beams. The essence of the following experimental work is to write programs to measure the required times and to graph the resulting data.

Each of the four light beams which traverse the glycerine column have several elements. An LED light source activates a cadmium sulfide photo-resistor whose resistance changes when light shines upon it. To sense this resistance change and to convert it into a digital signal suitable for computer processing, a voltage comparator circuit is used.

An LED is a small solid state light bulb which requires about 10 mA of current and 1.5 V to operate. A higher voltage source is generally used

Fig. 7.6. Viscometer apparatus.

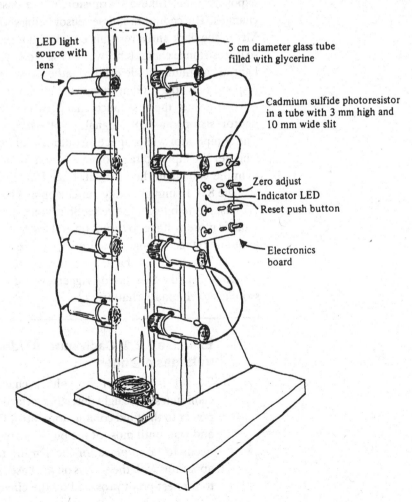

LED light source with lens

5 cm diameter glass tube filled with glycerine

Cadmium sulfide photoresistor in a tube with 3 mm high and 10 mm wide slit

Zero adjust
Indicator LED
Reset push button

Electronics board

Fig. 7.7. Optical position sensor circuit.

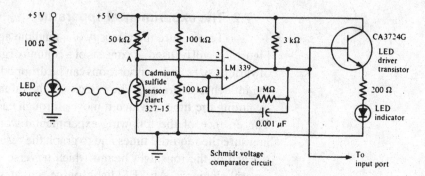

together with a current limiting resistor in series as shown in Figure 7.7. An LED passes current in only one direction so it is important that it be connected with the correct polarity.

Cadmium sulfide photoresistors are used in many cameras to compute the exposure time. Like a thermistor, it is a passive device whose resistance changes. The cadmium sulfide sensor being shown has a resistance of over 20 MΩ in the dark and a resistance in the hundreds of ohms in bright sunlight; so its resistance changes by over 100 000 to 1. Although it is quite sensitive to light, a cadmium sulfide cell is a rather slow device; it takes about 30 ms to respond fully to a sudden change in light level.

To translate the resistance change which the light beam induces in the photoresistor to a digital signal, a voltage comparator is used (Figure 7.7). The comparator will produce an output of either 5 V or 0 V depending upon whether the input voltage to the + input of the device is greater than or less than the voltage to the − input. Each LM339 has four such comparators in a single 8 × 15 mm integrated circuit chip. The comparator circuit in Figure 7.7 has a little bit of positive feedback incorporated to give latching action; it takes more voltage to turn it on and less voltage to turn it off than just the minute voltage change required to make the comparator switch. This hysteresis is similar to that used in the temperature controller of Chapter 2. The circuit is called a Schmidt trigger and is used frequently with mechanical switches to eliminate chattering.

Exercise 7.2.1 Cadmium sulfide cell resistance and voltages changes

To get a feeling of the voltage changes being registered by the cadmium sulfide light detectors, attach the wires and turn on the 5 V power to the fluid column apparatus. Fill the column with glycerine and wait until most of the bubbles are gone. The glycerine column needs to be in place for the sensors to focus correctly. Level the apparatus with the screws on the base. Attach an oscilloscope probe to the test point provided on the circuit board and put the oscilloscope in the free running mode with a sensitivity of 1 V/div. This point is the hot (not ground) side of the cadmium sulfide cell (point A, Figure 7.7). The 50 kΩ potentiometer which is in series with the photoresistor should be set so that the voltage at A is about one half the supply voltage, ie, 2.5 V. Break the light beam with a small piece of paper and note the voltage change which occurs. Move the paper across the light beam as fast as you can to get an idea of the minimum response time of the cadmium sulfide cell. Moving the paper vertically will probably give a faster response since the entry slit on the front of the tube is about 3 mm high and about 10 mm wide.

To set the potentiometer level for the experiment, turn it so that the LED goes off, then the other way until it just goes on. Test the setting by dropping a medium sized ball.

7.3 Using the 8253 timer

Even though the data taking rate in this experiment is modest by most standards, the timing methods used thus far are not precise enough. The DOS TOD (TimeOfDay) timer has a tock rate of about 55 ms or 1/20 of a second. To make a rough estimate of the time scale needed for the apparatus, assume the maximum velocity of fall to be reckoned with is about 0.3 m/s; this corresponds to the ball falling from the top of the column to the bottom (about 0.6 m) in about 2 s. The light beams have an effective width of about 3 mm; thus the computer should be able to record an instant of time t with a resolution of $t =$ distance/velocity $= 0.003/0.3 = 0.01$ s. This is five times faster than the TOD clock tocks.

Exercise 7.3.1 Speed of a sphere in air

With the apparatus described above, estimate the time resolution needed to measure the speed of a sphere falling through air.

Fortunately, the IBM-PC has an internal timer which runs at a much faster rate. In fact, it is that rate divided by 65 536 which gives the 55 ms TOD tock rate. The counting is done in a 8253 timer chip (see Appendix I for the data sheet on this chip). The chip has a 1.19 MHz clock rate as its input and has three 16-bit counters. We will be using timer 0 which is used by DOS for the TOD counter. Timer 1 is used for keeping the memory refreshed and timer 2 is used to generate tones on the speaker.

Like the 8255 digital I/O chip, the 8253 chip functions are controlled by registers. When the computer is started, timer 0 is initialized to count down repeatedly from $FFFF to zero and put out a square wave on its output wire

(OUT0). This is timer mode 3. The square wave will have a period of 65 536/1.19 MHz which is about 55 ms. OUT0 is then used as the TOD tock rate. The current count in timer 0 can be read by sending a control word to the timer and reading an eight-bit register twice. The control word tells the chip to read the current count and store it in a register. The counter continues to count down so that no ticks are lost but the register contains a static value. Then on two successive reads, the chip will send the low byte and then the high byte of the frozen (also called latched) count. The following Pascal statements illustrate the process:

```
Port[$43] := $00;              {$43 is the timer control register}
                               {putting 0 there freezes the count in}
                               {timer 0 and prepares the chip to send}
                               {the count bytes}
TimerLow  := Port[$40];        {$40 is the register for timer 0}
                               {the low byte comes first}
TimerHigh := Port[$40];        {now the high byte is read}
TimeTicks := Swap(TimerHigh) + TimerLow;
                               {Swap is a trick to get the high byte}
                               {in the right place easily, it could}
                               {have been written TimerHigh∗256}
```

Since the time for one tick is known ($0.838096\ \mu s$), successive access to the timer can then be used to time events up to about 55 ms long. For longer times the timer 0 and the TOD can be read. However, there is a problem with the way the timer works in mode 3 (the default mode). In order to produce square waves on the output, the chip counts down by twos and inverts the signal on OUT0 each time the count gets to zero. This makes it easy to get a square wave with exactly equal high and low times but makes the count which is read ambiguous. That is, it is not possible to tell whether the latched count was from the first half or second half of the cycle.

The remedy for this problem is to change the mode of timer 0. Mode 2 counts down by ones and produces a short pulse each time the count reaches zero (see the description in Appendix I). The TOD counter does not care whether the signal it gets is a square wave or a pulse so it will function the same in either mode. Since the count down is by ones, the latched count will be unique for each pulse.

To put the timer in mode 2 the following Pascal statements are used:

```
Port[$43] := $34;       {tell the chip that timer 0 is mode 2}
                        {and binary count, prepare chip for}
                        {initial count low byte first}
Port[$40] := $00;
Port[$40] := $00;       {start count at $0000, it will then}
                        {count $FFFF, $FFFE, . . .}
```

A careful reading of the control word format in the data sheets will show why $34 (%0011 0100) was used as the control word.

Exercise 7.3.2 Using the 8253 timer 0

(*a*) Write a Pascal program which initializes the 8253 timer 0 to mode 2 and then periodically reads the count and displays it on the screen.

(*b*) Add program steps so that the first two bytes of the TOD counter are also read and printed.

(*c*) Add steps so that the time in seconds is displayed. Since Pascal 16-bit integers give positive and negative numbers, the timer 0 and TOD counts must be handled carefully when converting to seconds.

By watching the program of Exercise 7.3.2(*b*) closely at the time the 8253 timer nears zero, you may be able to notice that a reading error can occur. It takes some time for the program to read the Timer 0 count and then read the TOD count. If the Timer 0 count reaches zero after it is latched for a read but before the TOD is read, the TOD will be incremented and be one count ahead of its value at the time of latching Timer 0. This will put the time 55 ms off.

If a way can be found to keep the TOD from incrementing until after it has been read, the problem would be solved. Fortunately this is easy to do. The TOD is updated by a process called interrupts. When an interrupt is signalled, the CPU stops in the middle of the program it was executing and executes a routine designated for that interrupt. Then it returns to the interrupted program as if nothing happened. Chapter 8 will cover this in much more detail. The OUT0 wire of timer 0 is connected to a wire which signals interrupts; so, every 55 ms an interrupt is generated. When that occurs the computer goes to a program which increments the TOD. This is happening all the time the computer is turned on.

There is a machine language instruction (CLI) which tells the 8088 to put any interrupts on hold and another (STI) which tells it to pay attention to them. It is like having a HOLD button on a telephone; the caller isn't disconnected but is just waiting. The instructions can be used for short periods of time to prevent the processor from being interrupted but can't be abused. For timer 0, if interrupts are turned off for longer than 55 ms then one tock will be lost since a second pulse will come in before the first has been serviced.

Exercise 7.3.3 Interrupt protection

Use Inline CLI and STI statements to protect the critical part of the code of the program you wrote for Exercise 7.3.2.

Exercise 7.3.4 Inline timer 0 access

So that the time is obtained as quickly as possible, rewrite the timer

0 and TOD access part of the program of Exercise 7.3.3 using Inline machine code. So that it can be used later, use a procedure which when called gets the time counts and converts them to a real number of seconds. Try to minimize the time that the routine takes. Test the speed of the routine by calling it twice in a row and displaying the time difference.

Exercise 7.3.5 Measuring ADC sample speed

Use the routines in Exercise 6.11.1 along with the timer 0 counter to determine the maximum speed the computer can convert and store numbers using the ADC (the maximum ADC sample rate). What is the maximum frequency signal which can be accurately sampled by this program?

7.4 Subtraction and addition

Turbo Pascal provides the number types real, integer, and byte for storage of data. The real type can represent very large and very small numbers to a precision determined by the number of bits allocated. However, real-number calculations take much longer to execute than integer calculations even with a math coprocessor helping. If speed is important and integers can be used, then integer calculations should be used. But integers have a limited range. If they are 16 bits long then they can only take values from $-32\,768$ to $32\,767$. This limits their usefulness considerably.

As an example, in the previous section there are two 16-bit numbers which come from the time counters. One from the TOD counter and one from timer 0. To make a routine which determines a time interval as quickly as possible, it would be most efficient if the two sets of 32 bits could be subtracted from each other rather than converting to reals and subtracting. (Remember that the timer 0 count needs to be inverted also.) The result would be in ticks instead of seconds but that conversion could be done at a later time when speed was not as important. Figure 7.8 shows a DEBUG listing of how double precision (32-bit) addition and subtraction is done in machine and assembly language.

The addition is started by a MOV to get the lower 16 bits of the first 32-bit word to the AX register. Then the AX register is added (ADD) to the lower bits of the second 32-bit word and stored in the AX register. AX is then saved in a third location. The higher 16 bits of the first word are then loaded into AX and added with carry (ADC) to the higher bits of the second word and moved to a third location. The first add was without the carry (ADD) because it is the first one done. Just as in adding two decimal numbers there is no carry into the first column. The carries between the 16 binary bits are

Fig. 7.8. Assembler listing of double precision addition and subtraction.

{The following does an addition of two 32-bit numbers $X + Y = Z$}

```
MOV  AX,[XLO]
ADD  AX,[YLO]
MOV  [ZLO],AX
MOV  AX,[XHI]
ADC  AX,[YHI]
MOV  [ZHI],AX
```

{The following does a subtraction of two 32-bit numbers $X - Y = Z$}

```
MOV  AX,[XLO]
SUB  AX,[YLO]
MOV  [ZLO],AX
MOV  AX,[XHI]
SBB  AX,[YHI]
MOV  [ZHI],AX
```

automatically taken care of by the adder in the CPU but the carry out of the 16th bit is stored in the carry bit of the Flag register. The carry bit thus indicates if the addition of the first 16 bits overflowed the storage capacity of 16 bits. The add with carry (ADC) will then automatically use the carry bit when adding the higher 16-bit numbers together. Since the ADC instruction also sets the carry flag, overflow from the higher bits could be tested by a branch instruction if necessary. Further chunks of 16-bit numbers could also be added, if larger numbers were needed.

The subtraction code of Figure 7.8 works in an analogous way except that the carry bit is now considered a borrow. The SUB instruction (like the ADD) ignores the borrow (carry) when subtracting the two numbers, but sets it to 0 if a borrow was not necessary and 1 if a borrow was necessary. The SBB (like the ADC) then subtracts the value of the borrow as well as the two numbers. It also sets the borrow for testing or for higher order bytes.

Exercise 7.4.1 Double precision integer subtraction

(a) Write and test a procedure which subtracts two 32-bit numbers and produces a 32-bit result. Use Inline statements. You may want to define a new data type

```
doubleint = record
                lo : integer;
                hi : integer;
            end;
```

If x is declared as doubleint then refer to x . lo in assembler by [x] and to x . hi by [x+1].

(b) Write and test a procedure which subtracts two sets of the timer 0 and TOD counts and produces a 32-bit result which is the number of ticks between two events. Remember that timer 0 is a count down timer and TOD is a count up counter.

(c) In Turbo Pascal 4.0 integers can be declared as integer (16 bits) or as longint (32 bits). (The other possible integer declarations are shortint, byte, and word but these do not concern us now.) So the above exercise may seem moot. However, consider the possible need to subtract two numbers represented by the two words from the TOD (16 bits each) and the Timer 0 word. That is a total of 48 bits. Write and test a procedure which subtracts two 48-bit integers and produces a 48-bit result which is the number of ticks between the two events. Use it in a program which outputs the exact time in ticks and microseconds between two successive key presses on the keyboard.

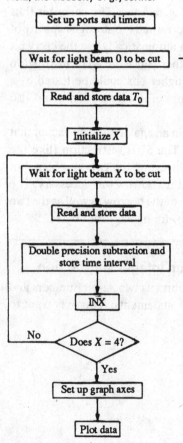

Fig. 7.9. Flow chart for Exercise 7.5.2, the viscosity of glycerine.

7.5　The viscometer

Exercise 7.5.1　Light beam sensing and timing

Connect the fluid column outputs to port C of the 8255 digital I/O chip. Write and test a program which tests for the first light beam to be cut and report the time at which this occurs. Test with a piece of paper interrupting the light beam.

Exercise 7.5.2　The viscosity of glycerine

(a) Write and test a program outlined by the flow chart in Figure 7.9 which waits for the subsequent light beams to be cut, measures the time interval from the cutting of the first beam and then plots the data on a graph.

Use your position vs. time plots to determine the terminal velocity and calculate the viscosity and Reynolds number for several balls of different diameters and densities. It is not necessary to do a least squares fit for each plot. Have the computer use two of the measured times to calculate the velocity and draw a line. You can check visually to make sure the other points fall along the line. If you input the diameter and mass of the ball, the computer can then calculate the viscosity (using Equation (7.1.5)) and Reynolds number (Equation (7.1.7)) and print them on the graph, too. Make

Table 7.1 *Typical diameters and masses of spheres*

Diameter (cm)	Mass (g)	Material
0.60	0.26	glass
1.31	2.7 ± 0.1	glass
1.575	5.2 ± 0.1	glass
0.09	4.1	lead
1.17	9.22	lead
1.45	17.8	lead
0.80	2.02	steel

several graphs with balls which are available to you. Some useful sizes are shown in Table 7.1.

(b) Compare your determinations of the viscosity with the value given in a reference book. Note that temperature and water content have a large effect on the viscosity of glycerine. (Exercises 7.5.4 and 7.5.5)

(c) Using the data for several balls, make a plot of the drag force on the ball (which equals the gravity minus buoyancy forces) vs. its terminal velocity times its radius (vr). Why is this plot significant? Can the transition to turbulence be seen?

(d) Will the timing part of your program work for a ball dropping in air (no glycerine)?

(e) Try replacing the glycerine in the column with water and repeating some of the measurements. Lead balls work the best in this case. Since the flow will be well into the turbulent regime ($Re \sim 400$), the viscosity cannot be accurately determined (why?). However, the drag coefficient C_d can be plotted vs. Re by assuming a value for the viscosity of water ($\mu = 0.010$ poise at 20 °C).

One experimental problem with this apparatus is that with larger diameter balls, the walls of the column interface with the flow and affect the motion of the ball. The viscosity value can be corrected with the following empirical formula (Dinsdale & Moore, *Viscosity and its Measurement*, Reinhold Publishing, New York, 1962):

$$\mu_{true} = \mu_{measured}[1 - 2.104(r/R) + 2.09(r/R)^2 - 0.95(r/R)^3]$$

where R is the radius of the column and r the radius of the ball.

Exercise 7.5.3 The wall effect

Correct the viscosity values obtained in Exercise 7.5.2 to account for the wall effect.

The temperature and water content of glycerine affect its viscosity greatly. The water content is particularly hard to control since glycerine absorbs water vapor from the air when it stands uncovered.

Exercise 7.5.4 Temperature variation of the viscosity of glycerine

The data shown in Table 7.2 taken from the *Handbook of Chemistry and Physics* and the *American Institute of Physics Handbook* show the temperature dependence of the viscosity of glycerine. Make a plot of viscosity vs. temperature. Suspecting an exponential dependence, now plot the natural logarithm of the viscosity vs. temperature and find the parameters and the model equation which give the best fit.

Table 7.2

Temperature (°C)	Viscosity (Pa s)
0	12.1
6	6.26
10	3.95
15	2.33
20	1.49
25	0.954
30	0.625

Exercise 7.5.5 The viscosity of aqueous solutions of glycerine

The data shown in Table 7.3 from the *Handbook of Chemistry and Physics* (Chemical Rubber Co., 52nd Edition, page D191) gives the relative viscosity of aqueous solutions of glycerol by percentage weight of glycerol.

(a) Plot these data to see the general behavior. Try both linear and log plots.

(b) Try fitting these data with the mixture formula:

$$\frac{1}{\mu} = \frac{P}{\mu_1} + \frac{1 - P}{\mu_2}$$

where P is the concentration of component 1 and μ_1 and μ_2 are the viscosities (μ_1 = glycerine, μ_2 = water).

(c) Try fitting these data with the Arrhenius formula (Dunstan & Thole, *The Viscosity of Liquids*, Longmans Green and Co., London, 1914).

$$\mu = \mu_1^P \mu_2^{1-P}$$

Table 7.3

% glycerol by weight	Relative viscosity (Viscosity/viscosity of water)
1	1.02
10	1.29
20	1.73
30	2.45
40	3.65
50	5.92
60	10.66
70	23.00
80	59.78
88	147.20
92	383.70
96	778.90
98	1177.00

(*d*) Try fitting a simple exponential to the data above 80% concentration.

8 Interrupts

Interrupts are an important capability of modern computers. They allow the processing of several independent tasks by the CPU. On large computers they allow multiuser and time sharing activities. On microprocessors they allow the running of a main program while periodically taking data or sending data to a slow device like a printer. Also computer start up, the keyboard, and TOD functions make use of the interrupt function.

In the discussion which follows, we will first trace the steps taken by the CPU when it receives an Interrupt Request (INTR) from other parts of the computer and then look into the ways we can cause interrupts to be generated and serviced.

8.1 Interrupts and the CPU

The interrupt sequence is similar to a jump to a subroutine except that it occurs when signalled by a wire leading to the CPU (INTR). When an interrupt signal is present on the INTR wire and the interrupt enable bit (IF) of the Flag register (see Figure 8.1) is 1, the CPU begins processing the interrupt. The interrupt enable bit is used to prevent the CPU from beginning to process the same interrupt again before it has completed the first one. Without it the computer would go into a continuous regression. The IF bit is set equal to 0 during an interrupt sequence and further interrupts are ignored until this bit is returned to 1. This can be done with the SEI instruction but is done automatically at the return from the interrupt service routine.

If the IF bit of the Flag register is 1, the CPU recognizes an INTR signal, and after completing the machine language instruction currently in process,

Fig. 8.1. 8088 Flag register bits (s = status, c = control).

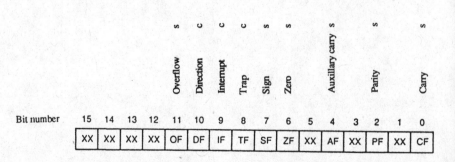

stores the Flag, CS, and IP registers on the stack. It is remarkable but necessarily true that if all registers of the CPU are restored to the values they had just before the interrupt occurred, the interrupted program will continue as if it had not been interrupted. (If only that could happen for an interrupted train of thought!) The CPU automatically stores the Flag, CS, and IP on the stack; it is the responsibility of the assembly language programmer to store and later restore any other registers which could be modified in the interrupt subroutine. Saving the CS and IP registers saves the location of the next instruction to be executed in the interrupted program.

After saving the Flag register on the stack, the IF flag is cleared so that the same interrupt cannot interrupt again leading to an infinite regression. The CPU then needs to be told where to find the machine code which is to be executed in response to that particular INTR. The 8088 uses a system of vectored interrupts. That is, there is a table of addresses (segment and offset) in the lowest section of memory (see the low memory map in Appendix C). Each possible INTR signal source has a corresponding entry in the table which is the address of the machine code to execute when that interrupt occurs. This machine code is called the Interrupt Service Routine (ISR). When an interrupt occurs the 8088 reads a byte from the data buss. The byte is the location in the vector table of the four-byte address to use. It then gets the address of the proper machine code from that vector location.

In the IBM-PC there are eight possible hardware interrupts (IRQ0–IRQ7) which are controlled by an 8259 interrupt controller. This device has control registers which can be programmed to enable or disable any of the eight interrupt signals connected to it. If an IRQ is signalled on a wire which is enabled it is passed on to the INTR wire of the 8088. Then the 8259 passes on the proper vector number in the table for that IRQ.

As an example we will study the operation of the TOD clock interrupt which we have been using for timing operations. The output of timer 0 of the 8253 timer chip is connected to IRQ0 of the 8259 chip. Every time the count in the timer 0 counter reaches zero (about every 55 ms), an IRQ signal is sent to the 8259. At start up time, the ROM routines have programmed the 8259 to accept IRQ0 signals and to send vector number 8 when the IRQ0 interrupt is signalled. Since entries in the vector table are four bytes long (segment and offset), vector 8 address occupies locations $0000:$0020 through $0000:$0023. The start up ROM routines also initiallized the address at vector 8 to $F000:$FEA5 which is an address in ROM. The ISR code at that address is shown in Figure 8.2.

The code begins by turning interrupts back on so that those needing more immediate attention can interrupt the TOD code. Then three registers are put on the stack (PUSH). These registers are used in the routine so they are saved here and restored at the end (POP). Next the segment ($0040) of the timer counters is placed in the DS register and the low word (TIMER_LOW at $006C) is incremented. (Notice that $0040:$006C is equivalent

Fig. 8.2. Timer interrupt code in BIOS from *IBM Technical Reference*, 1st edn, August 1981 (used by permission).

```
5689      ;-----------------------------------------------
5690      ; THIS ROUTINE HANDLES THE TIMER INTERRUPT FROM
5691      ; CHANNEL 0 OF THE 8253 TIMER.  INPUT FREQUENCY IS 1.19318 MHZ
5692      ; AND THE DIVISOR IS 65536, RESULTING IN APPROX. 18.2 INTERRUPTS
5693      ; EVERY SECOND.
5694      ;
5695      ; THE INTERRUPT HANDLER MAINTAINS A COUNT OF INTERRUPTS SINCE POWER
5696      ; ON TIME, WHICH MAY BE USED TO ESTABLISH TIME OF DAY.
5697      ; THE INTERRUPT HANDLER ALSO DECREMENTS THE MOTOR CONTROL COUNT
5698      ; OF THE DISKETTE, AND WHEN IT EXPIRES, WILL TURN OFF THE DISKETTE
5699      ; MOTOR, AND RESET THE MOTOR RUNNING FLAGS
5700      ; THE INTERRUPT HANDLER WILL ALSO INVOKE A USER ROUTINE THROUGH INTERRUPT
5701      ; 1CH AT EVERY TIME TICK.  THE USER MUST CODE A ROUTINE AND PLACE THE
5702      ; CORRECT ADDRESS IN THE VECTOR TABLE.
5703      ;-----------------------------------------------
FEA5            5704    TIMER_INT    PROC    FAR
FEA5 FB         5705            STI                        ; INTERRUPTS BACK ON
FEA6 1E         5706            PUSH    DS
FEA7 50         5707            PUSH    AX
FEA8 52         5708            PUSH    DX                 ; SAVE MACHINE STATE
FEA9 B84000   R 5709            MOV     AX,DATA
FEAC 8ED8       5710            MOV     DS,AX              ; ESTABLISH ADDRESSABILITY
FEAE FF066C00 R 5711            INC     TIMER_LOW          ; INCREMENT TIME
FEB2 7504       5712            JNZ     T4                 ; TEST_DAY
FEB4 FF066E00 R 5713            INC     TIMER_HIGH         ; INCREMENT HIGH WORD OF TIME
FEB8            5714    T4:                                ; TEST_DAY

FEB8 833E6E0018 R 5715          CMP     TIMER_HIGH,018H  TEST FOR COUNT EQUALLING 24 HOURS
FEBD 7519       5716            JNZ     T5                 ; DISKETTE_CTL
FEBF 813E6C00B000 R 5717        CMP     TIMER_LOW,0B0H
FEC5 7511       5718            JNZ     T5                 ; DISKETTE_CTL
                5719
                5720    ;------ TIMER HAS GONE 24 HOURS
                5721
FEC7 C7066E000000 R 5722        MOV     TIMER_HIGH,0
FECD C7066C000000 R 5723        MOV     TIMER_LOW,0
FED3 C606700001 R 5724          MOV     TIMER_OFL,1
                5725
                5726    ;------ TEST FOR DISKETTE TIME OUT
                5727
FED8            5728    T5:                                ; DISKETTE_CTL
FED8 FE0E4000 R 5729            DEC     MOTOR_COUNT
FEDC 750B       5730            JNZ     T6                 ; RETURN IF COUNT NOT OUT
FEDE 80263F00F0 R 5731          AND     MOTOR_STATUS,0F0H      ; TURN OFF MOTOR RUNNING BITS
FEE3 B00C       5732            MOV     AL,0CH
FEE5 BAF203     5733            MOV     DX,03F2H               ; FDC CTL PORT
FEE8 EE         5734            OUT.    DX,AL                  ; TURN OFF THE MOTOR
                5735
FEE9            5736    T6:                                ; TIMER_RET:
FEE9 CD1C       5737            INT     1CH                ; TRANSFER CONTROL TO A USER ROUTINE
FEEB B020       5738            MOV     AL,EOI
FEED E620       5739            OUT     020H,AL            ; END OF INTERRUPT TO 8259
FEEF 5A         5740            POP     DX
FEF0 58         5741            POP     AX
FEF1 1F         5742            POP     DS                 ; RESET MACHINE STATE
FEF2 CF         5743            IRET                       ; RETURN FROM INTERRUPT
                5744    TIMER_INT    ENDP
```

to $0000:$046C$.) If the low word overflows then the high word (TIMER_HIGH) is incremented. Then the high and low bytes are compared to the count for a complete 24 hours. If that has occurred, the counters are reset to 0 and the 24 hour roll over byte (TIMER_OFL) is set to 1. The next portion of the code (at T5) is to control the motor of the diskette drive. Then at T6 the software interrupt INT 1CH is invoked.

A software interrupt executes just like a hardware interrupt except the source of the IRQ is not a voltage on a wire but a machine instruction, INT. The number following the instruction is the interrupt number in the vector table to use. BIOS (ie, the Basic Input and Output System in ROM) makes use of many software interrupts. This gives the BIOS programmers much more flexibility. If the ROM location of a BIOS function (like putting a character on the screen) changes due to changes in the ROM program, only the initialization vector of the function in the vector table needs to be changed to the new location in order for software which uses that function to operate correctly. DOS also uses software interrupts for its services.

The INT 1CH instruction is put here to allow a programmer to have an ISR run every time this ISR is run. The address of the secondary ISR would go in the vector table at vector $1C. Initially the vector at that address is set to a simple IRET instruction so that nothing is done.

The next section of the ISR sends an EndOfInterrupt (EOI) command to the 8259. This must be done each time an interrupt is invoked through the 8259 so that the controller knows that the interrupt has been attended to. The command given here (sending $20 to port $20) is a general EOI which signals the completion of whichever interrupt is currently interrupting the CPU. This is the normal way of signalling the EOI to the 8259 controller. Remember this is only for hardware interrupts. Software interrupts aren't controlled by the 8259 so no EOI is sent for them.

The ISR ends with the restoration of the registers used (POP) and the IRET instruction which signals the CPU of the end of the interrupt routine. The CPU will then get the IP, CS, and Flag registers off the stack and proceed with the interrupted program. Notice because of the way interrupts work, an interrupt can interrupt an ISR of another interrupt. Also because of the way the 8259 works, several interrupts can be waiting to be serviced while one ISR is running. When that one is done then the next highest priority interrupt is invoked.

Turbo Pascal has several statements which allow interrupt procedures to be written easily. The Interrupt; declaration informs the compiler that it should insert register save instructions at the beginning of the procedure and register restore instructions at the end. The compiler also puts an IRET instruction at the end to replace the usual RET of a procedure. (In Turbo Pascal 3.0 you needed to put these in yourself.) Also provided are the GetIntVec and SetIntVec procedures to set the addresses in the vector table. Note that Turbo Pascal's input and output functions or DOS functions cannot be used in an interrupt routine since they are not reentrant. This means that if they are interrupted while executing and then used in the interrupt routine (reentered), they will not work.

Exercise 8.1.1 Using the TOD software interrupt

As stated above, the TOD ISR invokes software interrupt $1C when executed. Write and run the program below which will set up an ISR to beep every 32nd timer count. An EOI isn't sent to the 8259 because the timer ISR takes care of the hardware interrupt. Notice that the bell keeps ringing even when the program is stopped. Do not proceed until you have rebooted the computer to clear the interrupt.

```
program BellIRQ;
{rings terminal bell each 32nd time IRQ0 is activated}
{NOTE: reboot the computer after running this program!}
{else computer will freeze or bomb}

uses dos,                         {for interrupt statements}
     crt;                         {for Sound, NoSound}

procedure Beep;
{sound the speaker}
begin
    Sound(440);
    Delay(100);
    NoSound;
end;

procedure TimerTockISR( Flags,CS,IP,AX,BX,CX,DX,SI,DI,DS,ES,
BP : word);
{compares current count to determine if on even 32nd}
{parameters must be of the form above, see Turbo manual}

interrupt;                        {designate as an interrupt procedure}
                                  {see text explanation}

begin
If (MemW[$0040 : $006C] and $001F = 0 then
    Beep;
end;

var
    x : real;

begin
    SetIntVec( $1C, @TimerTockISR);    {set new vector, @ indicates}
                                       {the address of the procedure,}
                                       {that is a type 'pointer'}
    {do something to show that the IRQ operates independently}
    Repeat
        write('Enter a number(0 to stop) : ');
        readln( x);
        writeln('Sqr(x)= ',Sqr(x) : 15 : 4);
    until x = 0.0;

end.
```

The main part of the program starts by changing the $1C interrupt vector to point to the program. The SetIntVec; procedure uses CLI and STI instructions around the code which changes the vector because the program (and the computer) would bomb (or freeze) if a $1C interrupt occurred after only part of the address had been changed.

The ISR is still activated even after the program has stopped because it is still in memory and the vector still points to it. If another program overwrites the ISR, the vector will point to the middle of anything and most likely the computer will bomb. This will happen if you compile another program or even exit the Turbo system.

There are two remedies for this problem. One is to reset the vector of the interrupt to the original address before exiting a program which changes it. The other is to use a capability of the DOS operating system called Terminate-but-Stay-Resident. This instructs DOS to keep the code in memory permanently. DOS will then not overwrite the code with another program. The details of this function are beyond the scope of this text.

Exercise 8.1.2 Resetting the TOD interrupt

Modify the program of Exercise 8.1.1 so that the original address in the $1C vector is saved at the start and restored at the end of the program so that the ISR of the program is turned off and the computer does not need to be rebooted. Also make the interval between bells 64 timer tocks long.

8.2 Writing a hardware interrupt ISR

The John Bell Engineering board which has the 8255 PPI chips also has the means for generating interrupts and clock pulses. The circuit which will be used is shown in Figure 8.3. The 1 Hz clock is connected to a circuit

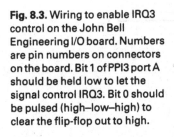

Fig. 8.3. Wiring to enable IRQ3 control on the John Bell Engineering I/O board. Numbers are pin numbers on connectors on the board. Bit 1 of PPI3 port A should be held low to let the signal control IRQ3. Bit 0 should be pulsed (high–low–high) to clear the flip-flop out to high.

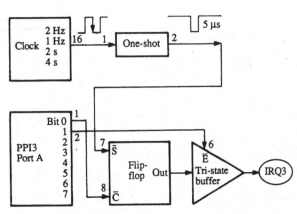

Fig. 8.4. OCW format for INTEL 8259 interrupt controller from data sheets (used by permission).

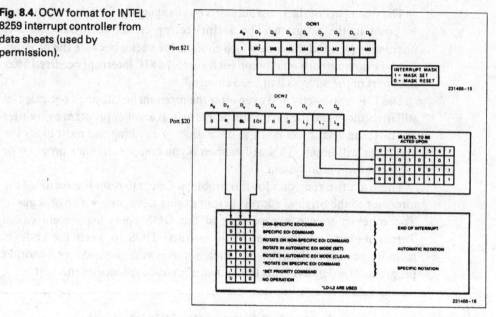

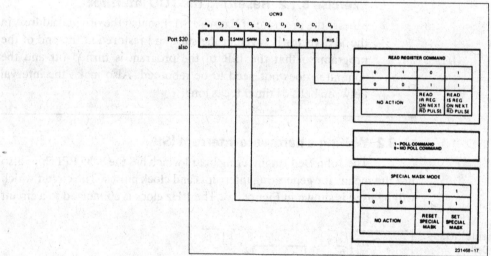

called a One-shot which outputs a short low pulse each time a downward going edge is input. The pulse goes to a flip-flop which also has connections to port A of PPI #3. The signal then goes through a tri-state buffer to IRQ3. This interrupt is sometimes used for the COM2 port but we will use it for our own purposes. This wiring is done on the I/O board inside the computer.

The flip-flop and the tri-state buffer allow us to control the generated interrupt. Assume the output of the flip-flop is low and the C input is high. When the 1 Hz signal goes low, a short (5 μs) low pulse is sent by the One-Shot to the $\overline{S}$ input of the flip-flop. The output of the flip-flop will then go high. It will stay high regardless of what the 1 Hz clock does until the $\overline{C}$ input gets a low pulse from bit 0 of the PPI #3 port A under program control. Then the output will go low until the next low on the 1 Hz clock. Thus the flip-flop is a one-bit memory cell; it has a high output if $\overline{S}$ (Set negative true) has pulsed low since the last time $\overline{C}$ (Clear negative true) has pulsed low. The tri-state buffer is just a gate. It has no output (high impedance) until the $\overline{E}$ (Enable negative true) input goes low. Then it just passes the high or low input through to the output. Thus the $\overline{E}$ input controls whether the IRQ3 is activated by the preceding circuit.

Writing a program which uses IRQ3 is quite similar to the program written in Exercise 8.1.2 to use the software interrupt of the timer. The main difference is that the 8255 PPI #3 must be programmed and the EOI signal needs to go to the 8259 interrupt controller at the end of the ISR. The 8259 also needs to be instructed to accept IRQ3 interrupts. This is done using the Operation Command Word (OCW) shown in Figure 8.4.

The OCW has two registers which are significant for us. At port address $20 is the OCW2 register which is used for the EOI command to signal the end of a hardware interrupt. As shown in the diagram, writing %001 into bits 7,6,5 of the register signals the end of the most recent IRQ. The OCW1 register at $21 controls which hardware interrupt is enabled (active). Placing a 1 in the corresponding bit disables the IRQ (mask set) and a 0 enables it (mask reset). Each bit corresponds to a particular hardware interrupt source. Thus the statement Port[$21] := Port[$21] AND $F7; will enable the IRQ3 interrupt wire leaving the other bits as they were.

Exercise 8.2.1 JBE 1 Hz interrupts

(a) Write a program which enables the 1 Hz clock to generate interrupts on IRQ3 and which responds to those interrupts by a beep. Use the structure of the program of Exercise 8.1.2 but keep the following ideas in mind.

(1) When starting the program, port A of PPI #3 needs to be set as an output and line A1 (Ebar) to zero so that the tri-state buffer is active (lets the signal through). Line A0 should be set from 1 to 0 and back to 1 so that the flip-flop is cleared at the start.

(2) As before the address of the ISR should be put in the vector location for IRQ3 (this is vector number 11.=$0B).

(3) In the ISR the flip-flop needs to be reset (and thus the IRQ3 signal cleared) by toggling the $\bar{C}$ (A0) line so that the next 1 Hz pulse can interrupt again.

(4) Also, the 8259 controller OCW needs to have the proper bit cleared thus enabling the interrupt.

(5) Before exiting the ISR the 8259 interrupt controller needs to be signalled that the interrupt has been taken care of; that is, send an EOI command like the ROM TIMER_INT procedure did in Section 8.1.

(b) If the $\bar{S}$ wire is available at the protoboard, try generating the interrupt using a push button instead of the 1 Hz timer. Use the circuit of Figure 6.8 and connect it to the $\bar{S}$ input of the flip-flop.

8.3 Serial data communication and the 8251 UART

In the exercises you have done, transmission of data to the computer has been direct. The sensors have been connected to the ADC or PPI which are inside the PC and connected to the internal buss. This is not always the case. Many newer instruments have means of gathering and storing digital data themselves. To analyze the data, they are transmitted along a cable between the instrument and computer. The methods used for this communication can be split into two broad groups: serial and parallel.

Serial data are transmitted one bit at a time. Each bit follows the previous one after a preset time interval has passed. This interval must be known to the receiver so that it can synchronize its timing with the transmitter. There are several hardware standards which are used for serial transmission. By far the most widespread is the RS-232C standard. It is used for slow to moderate speed communication (110–19 200 bits per second or 'baud') over distances of up to 300 m. Most terminals connected to multiuser computer systems use this standard as do many printers and plotters. At its minimum only two wires are needed: a ground and a signal wire. Since the standard requires that data only go one way on the signal wires, this minimal system would be good only for devices like printers. Most of the time another wire is added to provide two way communication and other wires to provide control signals. The RS-232C standard is also used to communicate with a modem which is a device that transmits and receives serial data over the telephone lines. A data rate of 300 or 1200 baud is commonly used.

Figure 8.5 shows how an ASCII character 'K' would be sent using the RS-232C protocol. (A summary of the ASCII code for encoding characters is shown in Figure 8.6.) The start bit signals the beginning of a data word. It is followed by 4–8 data bits. Then sometimes a parity bit is included which is used for error checking. At the end are one or two stop bits. The number of

Fig. 8.5. Serial transmission of an ASCII 'K' character. 'K' = Binary 01001011. The time for one bit is 1/BAUD rate.

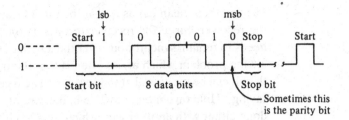

Fig. 8.6. ASCII Code. (Reproduced with permission from American National Standard X3.4–1986, copyright 1986 by the American National Standards Institute. Copies may be purchased from the American National Standards Institute, 1430, Broadway, New York, New York 10018.)

b7					0	0	0	0	1	1	1	1
b6					0	0	1	1	0	0	1	1
b5					0	1	0	1	0	1	0	1
b4	b3	b2	b1		0	1	2	3	4	5	6	7
0	0	0	0	0	NUL	DLE	SP	0	@	P	`	p
0	0	0	1	1	SOH	DC1	!	1	A	Q	a	q
0	0	1	0	2	STX	DC2	"	2	B	R	b	r
0	0	1	1	3	ETX	DC3	#	3	C	S	c	s
0	1	0	0	4	EOT	DC4	$	4	D	T	d	t
0	1	0	1	5	ENQ	NAK	%	5	E	U	e	u
0	1	1	0	6	ACK	SYN	&	6	F	V	f	v
0	1	1	1	7	BEL	ETB	'	7	G	W	g	w
1	0	0	0	8	BS	CAN	(	8	H	X	h	x
1	0	0	1	9	HT	EM	)	9	I	Y	i	y
1	0	1	0	10	LF	SUB	*	:	J	Z	j	z
1	0	1	1	11	VT	ESC	+	;	K	[	k	{
1	1	0	0	12	FF	FS	,	<	L	\	l		
1	1	0	1	13	CR	GS	−	=	M	]	m	}
1	1	1	0	14	SO	RS	.	>	N	^	n	~
1	1	1	1	15	SI	US	/	?	O	_	o	DEL

bits and their meanings as well as the rate of transmission must be known at the receiver. Since the receiver restarts its timing at each start bit, it only needs to remain synchronous over the length of the data word.

One problem which arises often is that the transmitter sends data faster than it can be processed at the receiver. The receiver needs to have a way of saying, 'Hold on a moment while I take care of what I already have.' This is done either with another wire which signals a hold or in software by having the receiver transmit characters to signal the transmitter. Most commonly the ASCII character 19 (Control S or XOFF) is HOLD and 17 (Control Q or XON) is GO. The transmission becomes a game of RED LIGHT GREEN LIGHT.

The transmission and reception of serial data is usually done by a UART (Universal Asynchronous Receiver Transmitter). Once it knows the protocol of the data being sent, the UART takes care of the serial interface. It is used by addressing registers; those for the 8251 chip are shown in the data sheet of Appendix J. On transmission, it translates the byte in its data register to serial form and on reception, it translates the serial data into a byte in the data register. The control registers are used to set the protocol of the serial data, the interrupt registers are for interrupt control, and the status registers are used to signal data transmission and error conditions. The interrupt capability is often used in communication programs so that the computer need not continually monitor the UART status.

Setting the communication parameters is a two-step process. First the speed of transmission (the baud rate) is set by setting bit 7 of the Line Control Register (LCR) to 1. This sets DLAB which enables Divisor Latch to receive the baud rate divisor. Notice that the Divisor latch occupies the same memory locations as the Transmit/Receive Buffer and the IER (see page 218). When DLAB is set to 1 the data goes to the Divisor Latch and when it is 0 the data goes to the other registers. The Divisor values to use for various baud rates are shown on page 230.

The second step for establishing the serial protocol is to set the LCR bit 7 to 0 and the other bits to the state required by the protocol. For example, to set a protocol of eight data bits, one stop bit, and no parity, the number %0000 0011 = $03 is used.

Once the parameters are set, the UART will transmit any ASCII code put into the TX Buffer. Bit 5 of the Line Status Register (LSR) indicates if the buffer is empty and another character can be sent to the UART. Bit 6 indicates if the transmitter has finished sending the character. Any character received will be available at the RX buffer. Bit 0 of the LSR indicates if a character has been received by the UART. Reading the RX buffer automatically resets this bit to zero and so it can be used to test if a new character has come in since the last read of RX. The other bits of the LSR indicate if there were any errors. They might occur if the wrong baud rate was being used or if the cable connecting the devices was faulty.

Exercise 8.3.1 Simple UART programming

(*a*) Write a procedure which initializes the UART to 1200 baud, eight data bits, one stop bit, and no parity. Write a program to test the UART transmission by sending the ASCII character 'K' repeatedly to the serial port. Using the oscilloscope, look at the resulting output on pin 2 of the serial connector and verify that it has the correct protocol and speed.

(*b*) Now wire pin 2 to pin 3 of the serial port so that the transmitted data is received by the same UART. Write a program which uses the same UART protocol and which continually loops through the following code:

(1) check if char has been typed at the keyboard (KeyPressed)

(2) if so, get char (ch : =ReadKey;) and send to the UART

(3) check for a character ready in the Receive buffer

(4) if so, get character and write to the screen.

In (2) be sure to check the TX buffer empty flag (THRE) before writing the next character into the buffer.

(*c*) When all is working well, use three wires to connect your computer to your neighbor's and talk back and forth.

8.4 Serial interrupt processing

The 8250 UART is also able to generate interrupts. In the PC the first serial port (COM1:) is connected to IRQ4. The UART can be programmed to send an interrupt signal for several different reasons. Table 24 of Appendix J summarizes the various possible sources. The Interrupt Enable Register (IER) is used to set which sources will cause interrupts. If several interrupt sources have been enabled, the Interrupt Identification Register (IIR) would be used by the ISR to determine which interrupt source caused the current interrupt to occur.

On the IBM-PC serial interface board the OUT2 wire from the UART has been connected to a tri-state buffer like the one on the John Bell Engineering board. In order for the interrupts generated by the UART to get to the IRQ4 wire the buffer must be activated by setting the OUT2 bit of the MODEM Control Register (MCR) to 1.

Exercise 8.4.1 Serial interrupt program

Write a program which uses the Receiver Data Available interrupt to receive data via the UART. The initialization routine should take the following steps:

(1) Set the serial port for 1200 baud, eight data, one stop, and no parity.

(2) Change the IRQ4 vector address to point to the ISR routine (SetIntVec).

(3) Hold interrupts ($FA is CLI) and enable Receiver Data Available interrupt in the UART.

(4) Enable the IRQ4 interrupts in the 8259 controller OCW.

(5) Clear RCR and LCR registers by reading them.

(6) Enable the tri-state buffer using the OUT2 bit of the MCR and start interrupts (STI).

The ISR should have the following steps:

(1) Get the data from the RCR buffer (this also clears the IRQ signal in the 8250).

(2) Send character to the screen (see the Chr function). Turbo Pascal I/O is not supposed to be re-entrant but see if it works anyhow.

(3) Send the general EOI to the 8259 controller.

The main program should have the following steps:

(1) Run the initialization procedure.

(2) Wait for a character from the keyboard (ReadKey).

(3) Transmit the character back out the UART (see the Ord function), also echo to the screen.

(4) If key is not '@' then loop back for another key else disable interrupts and stop program.

Test the program by wiring pin 2 to pin 3 of the serial port. When it is fully tested connect to your neighbor.

9 Other topics

9.1 Hardware for data acquisition and control

There are two styles of hardware for using a microcomputer to acquire data and control equipment. One is exemplified by the IBM-PC system you have used in the laboratory. The ADC, the DAC and the digital I/O cards are inside the computer and are under direct control of the microprocessor. They have control and data registers which are directly addressable via the buss. External devices (sensors, switches, etc) are connected to the cards. Creative programming can turn the computer into, for example, an oscilloscope (ADC and display) or a signal generator (DAC) as the laboratory exercises have shown.

Other buss systems are in use which, like the slots in the IBM-PC, allow a microprocessor to be connected to various data acquisition and control devices by simple board replacements. Some of the more widely used ones are S100, STDBUS, MULTIBUS, and QBUS.

The second style is to have a separate box next to the computer which has the ADCs, DACs, digital I/O lines and a programmed microprocessor controller. It communicates with the computer via a serial or parallel communication system (Section 9.2 has a description of a parallel system). The box takes care of the data acquisition and control while the computer is used to send control bytes to tell the box what to do and to receive the data for further processing. The limitation of this style is in the speed of communication to the computer and in the number of things the box has been preprogrammed to know how to do. However, some computers do not have card slots so that this style of data acquisition and control is the only possible choice.

9.2 Parallel data communication

In parallel transmission the data in one word are communicated simultaneously by having many wires connecting the transmitter and receiver. The data buss connecting various parts of the computer is one example; each bit of a data byte is stored in a memory location at the same time. To transmit and receive an eight-bit byte of data externally, eight wires are needed as well as several other wires, eg, a R/W wire, to control the direction and timing. A parallel hardware standard has been adopted for laboratory instrumentation which is called IEEE–488. Although the communication distance is limited to a total of 20 m, it can have up to 16 devices simultaneously connected and can transmit data at speeds up to 1 000 000 bytes per second. Many laboratory instruments now have options which allow connection to this buss.

In the IEEE–488 cable there are a total of 24 wires. Eight of these are ground wires which help to increase the noise immunity. There are eight data wires, three data transfer control wires and five management control wires. The devices on the buss can be designated as either talkers, listeners or controllers. There must be at least one controller which is usually a general purpose microcomputer. It manages the communication by using the management control wires to designate which devices should be listeners and which should be the talker. Only one talker is allowed at one time but the talker device can be changed at any time. For example, a printer would be a listener and a voltmeter would be a talker. Devices can also be active or inactive so, for example, the printer need not be printing all the time.

Communication of a byte of data is synchronized via a handshake mechanism using the three data transfer control wires. Figure 9.1 shows the sequence of signals to transmit one byte after the active talkers and listeners have been designated. Note that a low level indicates a true condition and a high level false. The sequence starts by each active listener letting the NRFD (Not Ready For Data) line go high (false) thus indicating that it is ready to receive data. Due to the open collector design of this signal wire interface, the signal does not go high until all of the listeners are ready. When the active talker sees the NRFD high it places the data on the data wires and signals

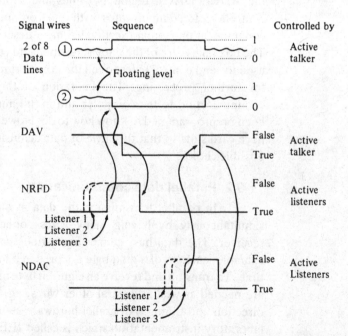

Fig. 9.1. IEEE-488 data transfer protocol. DAV is 'Data Valid', NRFD is 'Not Ready for Data', NDAC is 'Not Data Accepted'.

that the data is valid by dropping DAV (DAta Valid). The listeners then set NRFD LO and each store the data from the buss. As each completes that task, it lets the NDAC (Not Data ACcepted) signal go high indicating that the data has been stored. As with the NRFD, the NDAC wire does not go high until all the listeners have let it go. Thus the slowest listener active on the buss limits the speed of communication. The talker then sends DAV HI indicating that the data is not valid any longer and the listeners drop NDAC. The buss is then ready for the next byte transfer. This sequence is called a handshake since the data transfer takes place when both the transmitter and receiver have agreed (signalled) that they are ready.

The remainder of the wires in the buss are used for signals between the devices so that the talkers and listeners can be designated and so that the devices can signal emergency conditions. For example, if ATN (ATtentioN) is true it indicates to all the other devices that the controller wants to talk and that everyone else should listen. If SRQ (Service ReQuest) is true a listener is requesting to talk. The full protocol can be found by reading the interface documentation (Hewlett Packard calls it the GPIB interface) or by getting a description from the Institute of Electrical and Electronic Engineers.

Exercise 9.2.1 Parallel communication

Implement IEEE–488 protocol using the 8255 PPI. Assume that the eight data lines of Port B are connected to the data lines of the interface and that PC0 is connected to DAV, PC4 to NRFD, PC5 to NDAC. Also assume that the active talker is the computer. Write a program which will transfer 100 bytes of an array to the active listeners. Use the following outline:

(1) initialize the ports, set the data lines as inputs (floating temporarily), set DAV HI
(2) start loop of 100
(3) look for NRFD HI (all listeners ready)
(4) set data lines as output and put data on lines
(5) set DAV LO (signal data is valid)
(6) look for NDAC HI (data accepted by all listeners)
(7) set DAV HI and set data lines as inputs (floating again)
(8) loop back for next byte of data

9.3 DOS and BIOS function calls

One topic which has been flirted with but not completely introduced is that of function calls in the IBM-PC system. There are a certain number of functions which are commonly required to operate a computer and which are hardware dependent (such as sensing a keypress, putting a character on the screen, sending data to the disk drive). These functions are contained in the Basic Input and Output System (BIOS) and the Disk Operating System (DOS). The BIOS program is in ROM on board the computer with extensions which are loaded into RAM when a disk is booted. DOS is loaded into RAM when booting the computer. The functions in these programs are available for any software to use. They provide a common interface for programs so they don't individually need to take care of all possible hardware combinations.

These functions are accessed by means of software interrupts. When the computer is booted, the addresses of the functions are placed in their appropriate positions in the vector table. A program need only use a software interrupt (in Turbo Pascal Intr( ); procedure or in assembler the INT nn statement) to use the function. It doesn't need to know where the code is in memory, only the position in the vector table. A corollary is that a user written function can be used to replace a standard one simply by changing the vector in the table.

Table 9.1 shows the twelve ROM-BIOS functions available on IBM-PC type computers. Most programming languages (including Turbo Pascal) use these when executing standard language statements since they provide the basic access to the computer hardware. They can be accessed directly via the Intr( ); function if desired.

Most DOS functions are accessed via interrupt 33 with the AH register indicating which function to perform (see Table 9.2). These functions are not as universally used as the BIOS functions since they have limitations which some programmers wish to avoid. They can be accessed with the

Table 9.1 *Software interrupt ROM-BIOS services*

Interrupt		
Dec	Hex	Function
5	5	Print-screen
16	10	Video display
17	11	Equipment list
18	12	Memory size
19	13	Floppy disk
20	14	Communications
21	15	Cassette tape
22	16	Standard keyboard
23	17	Printer
24	18	ROM-BASIC
25	19	Bootstrap start-up
26	1A	Time and date

Table 9.2 *A few examples of DOS functions available using the DOS software interrupt (33 decimal). Many more are available*

AH register		
Dec	Hex	Function
0	0	End program
1	1	Keyboard input with echo
2	2	Display output
3	3	Serial input
4	4	Serial output
5	5	Printer output
8	8	Keyboard no echo
9	9	Display string
10	A	Buffered Keyboard input
15	F	Open disk file
16	10	Close disk file
20	14	Read sequential file record
21	15	Write sequential file record
22	16	Create a file
23	17	Rename a file
37	25	Set interrupt vector
41	29	Parse filename
42	2A	Get date
43	2B	Set date
44	2C	Get time
45	2D	Set time

MsDOS(); procedure in Turbo Pascal. For details on BIOS or DOS functions see Norton or the DOS Technical Reference listed in the bibliography.

9.4 EPROM and EEPROM

Erasable Programmable Read Only Memory (EPROM) is a cross between ROM (which can't be reprogrammed) and RAM (which forgets everything when the power is turned off). Like a ROM, an EPROM requires special equipment to write the data into its memory. It will not forget the data when the power is turned off; but unlike ROM, it can be erased by shining ultraviolet light through a quartz window in the top of the chip. Thus, programs can be developed by erasing and reprogramming improved versions on a single EPROM. In building experimental apparatus it is often convenient and economical to have a simple one board computer dedicated to doing a single task with a ROM or EPROM to store its program. An EEPROM is an Electrically Erasable Read Only memory which uses a high voltage to erase the memory instead of ultraviolet light.

9.5 Sensors and transducers

In the laboratory work in this book you have used only three kinds of sensors, a potentiometer, a thermistor, and a photoresistor, and two controllers, a stepping motor and a HEXFET switch. There are many other kinds of sensor, at least one for each physical parameter which is measured. A good physical understanding of the system to be measured is always the first step. Then, selection or design of the sensor can be done. Some generalized performance characteristics have been discussed in the sections on zero, first, and second-order systems. Understanding the physical and electrical basis of the sensor is also important. Please refer to the references for information on the wide variety available. Keep in mind that there is always room for invention.

9.6 Software for data acquisition and control

Of the large amount of software available for a particular microcomputer, there are two basic types: languages and application programs. The first are the primary tools with which a computer is programmed (eg, Pascal). The second are particular programs which have the computer perform specific tasks (eg, a graphing program). Both have their places in the use of the computer in the laboratory.

As in the work done in this book, most laboratory computers are programmed in the laboratory using a chosen language. Table 9.3 lists the more popular ones with some comments on their efficacy. A program in an interpreted language is executed as it is run whereas one in a compiled

Table 9.3 *Microcomputer languages*

Language	Comments
BASIC	Interpreted or compiled, common, easy to learn, slow unless compiled
Assembly	Compiled, most direct control of computer system, awkward
FORTRAN	Compiled, traditional for number-crunching analysis, has complex numbers!, awkward, frequently no direct memory access, libraries available
Pascal	Compiled, structured for easier programming
Ada	Like Pascal but US Department of Defense backing
Modula II	Like Pascal but corrects some weaknesses
FORTH	Threaded, can be extended by user, originated for data acquisition and control, somewhat awkward reverse polish constructs
C	Compiled, both low level and high level programming, structured, terse

language must be translated into machine code before it can be run. Be sure that the language has the capability of writing and reading absolute memory locations.

Most application programs for data acquisition which are available at this time are libraries of subroutines (or procedures or modules) which, when called, do specific tasks. For example, one subroutine would output a number to the DAC and another would get the time from the timer. The libraries are specific to the language and the hardware being used.

What really made microcomputers popular for the home and business were two applications programs: the word processor and the spread sheet. These are versatile programs dedicated to a specific need (such as writing) but general enough to encompass a variety of tasks within that need (such as letters, reports, lists). There are a few programs available which address the need for a generalized data acquisition, storage, analysis and graphing. As the business market saturates, it is to be expected that more and varied programs will be written for the scientist and engineer.

9.7 Where to go from here

A great deal of useful work can be done in the laboratory by applying the principles you have learned. For those interested, there are several areas of study which extend the topics discussed here. A laboratory course on digital and analog electronics would be useful in understanding sensors and their associated signal conditioning circuits as well as the electrical operation of the computer itself. An introductory course in signal processing and analysis would be useful for general data analysis. For those interested in process automation, a course in systems analysis would be helpful. To keep up on the latest hardware and software in this quickly changing field, consult trade journals. Also get on the mailing lists of suppliers. They will frequently send out product bulletins. But the best way to learn is the way you have learned in this laboratory, ie, by doing it.

Appendix A
Laboratory materials and sources

The following is a detailed description of the equipment used in the laboratory at Cornell University together with possible sources for these parts.

Each student work area (Figure A.1) has an IBM-PC computer with printer and data acquisition cards, a 5 V power supply, and an oscilloscope. As stated in Chapter 1, IBM-PC, XT, AT, or design copies of any of these can be used but make sure that the computer has a slot for the data acquisition board (Figure A.2) and that the computer runs the software used (eg, Turbo Pascal 4.0 and a graphing program, see Appendix B). IBM-PS/2 systems do not have the same architecture so are not suited for direct use with this book. PC-DOS or MS-DOS are equally usable as the operating system. The book was written with Turbo Pascal 4.0 or later in mind. Version 3.0 is usable with minor differences. The general user interface and the interrupt procedures being the most notable.

Note for AT class CPU users: The address 800($320) conflicts with the hard disk I/O address so place the I/O board base address at 544($220).

Fig. A.1. General setup in the laboratory. Computer, monitor, printer, oscilloscope and interface prototype board are all centralized. A stand for the printer which puts the printer over the oscilloscope is useful.

References to I/O board addresses should be changed throughout the book. For instance, in program ADCTest (p. 14) the ADCRegister address becomes 556 or 544+12.

We use a B+K Precision Model 1476A dual trace 10 MHz oscilloscope and a Power One model C5–6 power supply. Almost any oscilloscope will do

Fig. A.2. The John Bell Engineering Universal I/O card for the IBM-PC with cables to connect the interface prototype boad.

Fig. A.3. A view of the prototype boards where the interface cables terminate. The ADC cable is to the left and digital I/O cables to the right.

and the only specification which needs to be met on the power supply is that it has a 5 V output at 5 A. Look in the back of *BYTE* magazine or in surplus catalogs for good prices. We have tied the computer and power supply grounds together permanently so as to minimize grounding problems for the students.

The cables from the data acquisition boards are brought out to a proto-board (Figure A.3) where connections may be made easily. We find that the Super Strip (available through Digi-Key or Jameco) to be versatile.

Laboratory Apparatus

Potentiometer (Figure A.4)
Almost any will do in the resistance range of 100 Ω to 1 MΩ.

Thermistor calibration apparatus (Figure A.5)
A thermistor, thermometer, and heater resistor are mounted in an aluminum block about 2 cm × 2 cm × 2 cm size. The thermistor we use is a Fenwall GB34P2. Others may be substituted by adjusting the bias resistor depending on the room temperature resistance. Heat conductive grease is used in the holes so that the thermometer, thermistor and heater resistor make good

Fig. A.4. The push button and potentiometers.

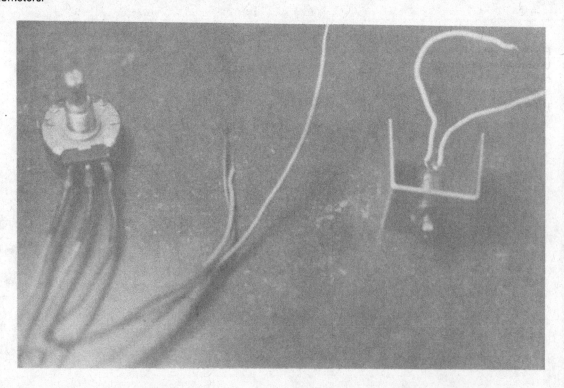

thermal contact with the block. The circuit used is shown in Figure 3.5. A standard laboratory mercury thermometer is used but others can be substituted. The HEXFET is an International Rectifier IRF 510. Almost any of that line can be substituted.

Stepping motor (Figure A.6)
The stepping motor apparatus consists of a stepping motor connected to a 200:1 gear box by a rubber sleeve and controlled by a UCN-4202A controller (Sprague Electric Co.) which is mounted on a protoboard. Figure 4.1 shows the circuit used. The stepping motor is a surplus item (A. W. Hayden Co. P/N B86138) which may be hard to find but the controller will work with Permanent Magnet stepping motors rated to 500 MA and 15 V. You may have to modify the wiring of the motor to suit the controller. *BYTE* magazine is a good place to look for surplus motors. The gearbox is from AST/SERVO Systems and again is a surplus item. The reduction ratio is not critical.

Fig. A.5. View of the thermistor calibration/temperature controller apparatus. The aluminum block at the top holds the heater resistor, the thermistor and the thermometer; all emplaced with conductive grease and some glue to hold them in place. The circuit is constructed on a piece of protoboard.

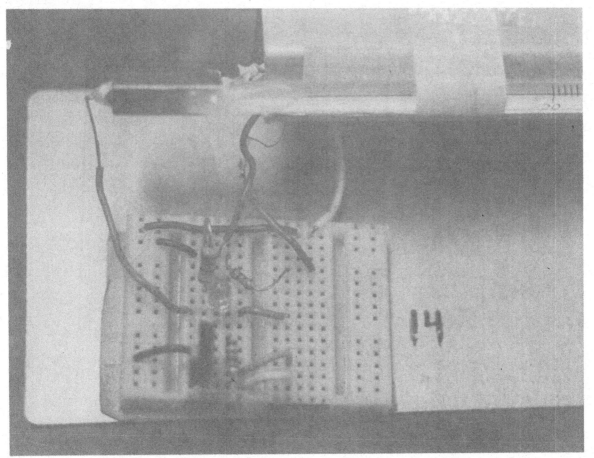

LED Output Counter (Figure A.3)

These are simple LEDs with 270 Ω resistors and a 74LS04 driver. The circuit (Figure 4.4) is constructed on the protoboard where the cables from 6522 VIA #2 are attached.

Heat Flow Apparatus (Figure A.7)

The apparatus for the heat flow experiment consists of a copper rod (#8 copper wire, 3.24 mm diameter) mounted vertically on an aluminium base as shown in Figure 5.3. An aluminum support runs parallel to the rod to support the wires to the heater resistor and thermistors. The $\frac{1}{4}$ W resistor is placed in a hole in the top of the rod. The thermistors (Fenwall GB32J2) are placed in small holes at 2.5 cm and 5 cm down from the resistor. Thermal grease is again used to ensure thermal contact. The standard amplifier circuit employed is shown in Figure 5.5. A protoboard is used to construct the circuit. The only special consideration is that the operational amplifier be able to run on 0–5 V supplies.

Fig. A.6. Stepping motor apparatus. The protractor is mounted to the left on the output shaft of the gearbox (center). The stepping motor (on the right) is mounted to the gearbox and coupled with a piece of rubber tubing. The circuit is constructed on a piece of protoboard.

Digital to Analog Converter (Figure A.3)

This circuit (Figure 6.7) is constructed on the protoboard where the LEDs and the John Bell Engineering card 8255 PPI cables are attached. The DAC used is a National Semiconductor DAC0808. Others may be substituted with some change in circuitry. The negative voltage necessary for running this chip can be obtained from the John Bell Engineering card buss connector or a voltage inverter.

Viscometer (Figure A.8)

As depicted in Figure 7.6, the viscometer apparatus consists of a glass tube about 5 cm in diameter with a rubber stopper at one end mounted in a wooden frame to which the detectors and electonics are attached. The frame may be leveled by means of the three screws at its base. The light source for the position sensors are green LEDs mounted in 1 cm aluminum tubes with a small focusing lens at one end. The light detectors are cadmium sulfide photoresistors (Claret 327–15) mounted in another 1 cm aluminum tube. In front of the sensor is a 3 mm high 10 mm wide slit cut in cardboard. The

Fig. A.7. Heat flow apparatus. The copper wire is secured to the aluminum base plate and has three holes drilled for mounting the heater resistor and the two thermistors. These are emplaced into the copper with thermal grease and secured with glue. Their leads are supported with a piece of aluminum. The circuit is constructed on a piece of protoboard attached to the base.

circuit for one of the sensors is shown in Figure 7.7. The balls used can be of a wide variety however the 'wall effect' becomes very evident for large ones. Table 7.1 shows some we have found useful.

Fig. A.8. Viscometer. A glass tube, the four positions sensors and the electronics are mounted on a wooden base which can be leveled by adjusting three screws.

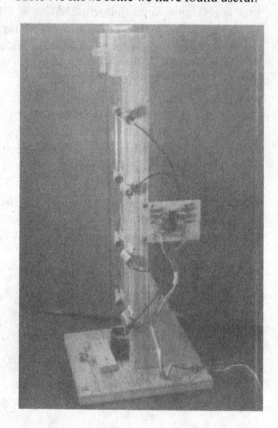

Addresses

John Bell Engineering
400 Oxford Way
Belmont, CA 94002

Borland International
4585 Scotts Valley Drive
Scotts Valley, CA 95066

Digi-Key
PO Box 677
Thief River Falls, MN 56701

Jameco Electronics
1355 Shoreway Road
Belmont, CA 94002

Sprague Electric Co.
115 Northeast Cutoff
Worcester, MA 01606

AST/SERVO Systems Inc
930 Broadway
Newark, NJ 07104

Kits of all the laboratory apparatus can be purchased from:
Bruce Thompson
30 Sodom Road
Ithaca, NY 14850
USA

Appendix B
Graphing programs and disk configuration

In addition to Turbo Pascal, a graphing program is needed which is capable of drawing engineering graphs on the screen and the printer. There are several ways to do this. One is to use the primitive graphics functions of Turbo Pascal itself to build a library of graphing procedures. This has the advantage of making the graphics functions available in the program generating the data but is a much larger task than it initially appears to be. The Turbo Graphics Toolbox available from Borland International provides some plotting features, but generally it concentrates on graphics rather than graphing which limits its usefulness.

Using a spread sheet program is another way of graphing data. It has the advantage of also allowing the data to be transformed before graphing. However, the graph formats are frequently limited (eg, a single set of abscissa values and no logarithmic scales) and a file of the data must be written before it is used.

There are a number of stand alone graphing programs on the market now which can make beautiful graphs. Some of them allow data manipulation as well. Please see the list of programs which follows.

A wish: Someone please make an engineering graphing program (memory resident?) which can be used from within a program to make quick graphs on the screen or printer of data and models.

Graphing program sources

A short list of stand alone type programs for engineering graphing.

Energraphics
Enertronics Research Inc.
St Louis, MO.

GenPlot
Michael O. Thompson
Material Science Dept
Cornell University
Ithaca, NY 14853

Grapher
Golden Software
PO Box 281
Golden, CO 80402

Plot
New Unit
Dewitt Building
Ithaca, NY 14850

2-D Graphics	SigmaPlot
Intex Solutions	Jandel Scientific
568 Washington St.	65 Koch Road
Wellesley, MA 02181	Corte Madera, CA 94925

System boot disk

The system boot disk for use with this book should have the following configuration so that it runs efficiently:

File	Type	Comment
COMMND.COM	DOS	Operating system file
FORMAT.COM	DOS	For formatting data disks
PRINT.COM	DOS	For printing files on the printer
GRAPHICS.COM	DOS	For printing graphics screen images
DEBUG.COM	DOS	For assembly language
CONFIG.SYS	DOS	To set system configuration
AUTOEXEC.BAT	DOS	To set system configuration
TURBO.EXE	Turbo	Turbo Pascal program
TURBO.TPL	Turbo	TP standard units
TURBO.HLP	Turbo	TP help, may be deleted if desired
TINST.EXE	Turbo	TP installation program
TURBO.TP	Turbo	TP installation data file created by TINST
GRAPH.TPU	Turbo	TP graph unit for real time graphing
PLOT.EXE	????	Your graphing program

Other programs can be added to this list if desired. If your plotting program is big, it may have to go on another disk. The plotting program may also require the presence of other files such as fonts and printer/screen drivers. Write protecting the disk may be a good idea.

AUTOEXEC.BAT should have the following minimum form:

```
PATH=A:\
PROMPT=$p $g
GRAPHICS
DATE
TIME
TURBO
```

where other date and time setting programs can be substituted for DATE and TIME. Running GRAPHICS installs the graphics screen dump in memory so it is always available. Of course other start up programs can be added if you wish.

CONFIG.SYS should have (if the computer memory is large enough):

```
FILES=16
BUFFERS=16
```

Appendix C
IBM-PC memory map

Figure C.1 shows how the address space of the IBM-PC is organized. On booting the computer DOS and BIOS extensions are loaded into RAM memory and executed. Some parts of these programs remain resident after they are run. One of their functions is to set up the proper addresses of interrupt subroutines in the vector table of low memory. Figure C.2 shows how the lowest memory addresses are organized. Programs are loaded beginning at the next free space in RAM memory. When a new program is run, it is loaded on top of the previous one (writing over it) unless the previous program has reserved space by means of the DOS Terminate but stay resident function (DOS function 39 or $27).

Fig. C.1. Memory map for IBM-PC.

Address		Function
Dec	Hex	
1024 K	FFFFF	Top of address space
	F0000	ROM-BIOS, ROM-BASIC, ROM-DIAGNOSTICS
	EFFFF	
		Other ROM or unused
	D0000	
	CFFFF	
		ROM-BIOS extensions
	C0000	
	BFFFF	
	BC000	Color/graphics display memory
	B8000	
	B4000	Monochrome display memory
	B0000	
	AFFFF	
		Display memory extension
	A0000	
640 K	9FFFF	
		RAM { DOS, resident programs, non resident programs
0 K	00000	

Fig. C.2. Low memory maps
(from *IBM-PC Hardware
Technical Manual*, 1st edn –
used by permission).

Table 30. Interrupt Vectors (0-7F)

ADDRESS HEX	INTERRUPT HEX	FUNCTION
0–3	0	Divide by Zero
4–7	1	Single step
8–B	2	Non-Maskable Interrupt (NMI)
C–F	3	Break Point Instruction ('CC'x)
10–13	4	Overflow
14–17	5	Print Screen
18–1F	6,7	Reserved
20–23	8	Timer (18.2 per second)
24–27	9	Keyboard Interrupt
28–37	A,B,C,D	Reserved
38–3B	E	Diskette Interrupt
3C–3F	F	Reserved
40–43	10	Video I/O Call
44–47	11	Equipment Check Call
48–4B	12	Memory Check Call
4C–4F	13	Diskette I/O Call
50–53	14	RS232 I/O Call
54–57	15	Cassette I/O Call
58–5B	16	Keyboard I/O Call
5C–5F	17	Printer I/O Call
60–63	18	ROM Basic Entry Code
64–67	19	Boot Strap Loader
68–6B	1A	Time of Day Call
6C–6F	1B	Get Control on Keyboard Break: Note 1
70–73	1C	Get Control on timer interrupt: Note 1
74–77	1D	Pointer to video initialization table: Note 2
78–7B	1E	Pointer to diskette parameter table: Note 2
7C–7F	1F	Pointer to table (1KB) for graphics character Generator for ASCII 128–255. Defaults to 0:0

Notes:　(1) Initialized at power up to point to an IRET Instruction.
　　　　 (2) Initialized at power up to point to tables in ROM.

ROM

8259 Hardware interrupts

Interrupt	IRQ	Use
8	0	Timer
9	1	Keyboard
A	2	
B	3	Serial (Com2)
C	4	Serial (Com1)
D	5	Printer (LPT2)
E	6	Diskette
F	7	Printer (LPT7)

Table 31. BASIC and DOS Reserved Interrupts (80-3FF)

ADDRESS HEX	INTERRUPT HEX	FUNCTION
80–83	20	DOS Program Terminate
84–87	21	DOS Function Call
88–8B	22	DOS Terminate Address
8C–8F	23	DOS CTRL—BRK Exit Address
90–93	24	DOS Fatal Error Vector
94–97	25	DOS Absolute Disk read
98–9B	26	DOS Absolute Disk write
9C–9F	27	DOS Terminate, Fix in Storage
A0–FF	28–3F	Reserved for DOS
100–1FF	40–7F	Not Used
200–217	80–85	Reserved By BASIC
218–3C3	86–F0	Used by BASIC Interpreter while BASIC is Running.
3C4–3FF	F1–FF	Not Used

Table 32. Reserved Memory Locations (400-5FF)

ADDRESS HEX	MODE	FUNCTION
400–48F	ROM BIOS	See BIOS Listing
490–4CF	DOS	Used by DOS Mode Command
4D0–4EF		Reserved
4F0–4FF		Reserved as Intra-Application Communication area for any application.
500–5FF		Reserved for DOS and BASIC
500	DOS	Print Screen status flag store. 0—Print screen not active or successful print screen operation. 1—Print screen in progress. 255—Error encountered during print screen operation.
504	DOS	Single drive mode status byte.
510–511	BASIC	BASIC's segment address store.
512–515	BASIC	Clock interrupt vector segment: offset store.
516–519	BASIC	Break key interrupt vector segment: offset store.
51A–51D	BASIC	Disk error interrupt vector segment: offset store.

Appendix D
Connections and logic of the ADC

To use apparatus intelligently it helps to understand what is going on inside; the discussion below focuses on giving some insight into what occurs when you do an analog conversion. As with most things, such discussion has many layers of increasing depth and detail. This discussion will go only one veneer down.

The analog to digital conversion is done on the John Bell Engineering Universal I/O board by an ADC 0817 IC which is connected to the address and data busses and to the RD (read) and WR (write) wires of the IBM-PC (Figure D.1). Port addresses 812–819 (assuming a base address of 800 decimal) are devoted to operation of the ADC.

There are basically three operations which can be done. Refer to Figure D.1 in the following. If a byte is written to any of the port addresses 812–815 (e.g., Port[812]:=n;), the address is decoded by the card circuitry and the CS4 wire is activated. The WR wire is also active since this is a write operation. Thus the data wires (D0–3) at positions ABCD on the multiplexer are activated. These four wires select by means of a binary code, the position of the switch so that one of 16 analog input channels is connected to the input of the converter section. Thus the lower four binary digits of the byte (n) are used to set the multiplexer channel number. Just after the multiplexer is switched, the converter is started since the WR and CS4 wires are also

Fig. D.1. ADC connections.

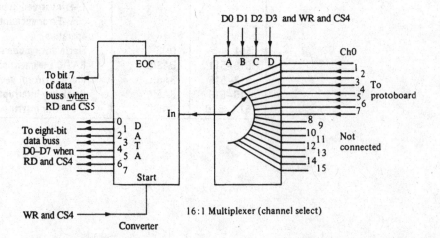

connected to the Start input of the converter section. And after it finishes, the data (the eight-bit representation of the voltage at the input) is available at the data output wires (0–7). There is no difference in using any of these addresses; Port[812]:=n; and Port[815]:=n; do exactly the same thing since they all activate the CS4 wire.

Reading the data on the output wires is the second operation which can be done on the ADC. This operation is done at the same addresses as above (eg, x:=Port[812];). The result of the last conversion is present on the output wires no matter which channel of the multiplexer is chosen. It is therefore necessary that the result of one conversion is read before the next is started. The data (0–7) is connected to the data buss when the RD and CS4 wires are active.

The third operation which can be programmed with the ADC is to check the status of the conversion. A read operation at port addresses 816–819 activates the RD and CS5 wires which connect the EOC (End of Conversion) output of the ADC to bit 7 of the data buss. Thus the read will return a byte whose bit 7 reflects the state of the EOC wire. If bit 7 is a 0, then the ADC is still busy converting the last channel addressed. If bit 7 is a 1, then the ADC has finished converting and the data is valid. It is not necessary to use this feature if a simple delay of sufficient duration (more than 200 μs) is used between starting the converter and reading the data. Section 6.11 has an exercise in using this feature.

Appendix E
8255 Programmable Peripheral Interface data sheets

Although cryptic, data sheets contain all of the detailed information about a particular device. But, be warned!, they are sometimes inaccurate due to typos and poor editing. The data sheets on the following pages of the 8255 PPI seem to be accurate.

8255A/8255A-5
PROGRAMMABLE PERIPHERAL INTERFACE

- MCS-85™ Compatible 8255A-5
- 24 Programmable I/O Pins
- Completely TTL Compatible
- Fully Compatible with Intel Microprocessor Families
- Improved Timing Characteristics

- Direct Bit Set/Reset Capability Easing Control Application Interface
- Reduces System Package Count
- Improved DC Driving Capability
- Available in EXPRESS
  — Standard Temperature Range
  — Extended Temperature Range
- 40 Pin DIP Package or 44 Lead PLCC

(See Intel Packaging: Order Number: 231369)

The Intel 8255A is a general purpose programmable I/O device designed for use with Intel microprocessors. It has 24 I/O pins which may be individually programmed in 2 groups of 12 and used in 3 major modes of operation. In the first mode (MODE 0), each group of 12 I/O pins may be programmed in sets of 4 to be input or output. In MODE 1, the second mode, each group may be programmed to have 8 lines of input or output. Of the remaining 4 pins, 3 are used for handshaking and interrupt control signals. The third mode of operation (MODE 2) is a bidirectional bus mode which uses 8 lines for a bidirectional bus, and 5 lines, borrowing one from the other group, for handshaking.

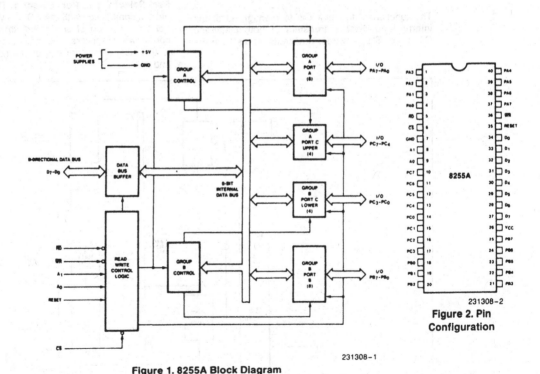

231308-1

Figure 1. 8255A Block Diagram

231308-2

Figure 2. Pin Configuration

8255A FUNCTIONAL DESCRIPTION

General

The 8255A is a programmable peripheral interface (PPI) device designed for use in Intel microcomputer systems. Its function is that of a general purpose I/O component to interface peripheral equipment to the microcomputer system bus. The functional configuration of the 8255A is programmed by the system software so that normally no external logic is necessary to interface peripheral devices or structures.

Data Bus Buffer

This 3-state bidirectional 8-bit buffer is used to interface the 8255A to the system data bus. Data is transmitted or received by the buffer upon execution of input or output instructions by the CPU. Control words and status information are also transferred through the data bus buffer.

Read/Write and Control Logic

The function of this block is to manage all of the internal and external transfers of both Data and Control or Status words. It accepts inputs from the CPU Address and Control busses and in turn, issues commands to both of the Control Groups.

(CS)

Chip Select. A "low" on this input pin enables the communication between the 8255A and the CPU.

(RD)

Read. A "low" on this input pin enables the 8255A to send the data or status information to the CPU on the data bus. In essence, it allows the CPU to "read from" the 8255A.

(WR)

Write. A "low" on this input pin enables the CPU to write data or control words into the 8255A.

(A₀ and A₁)

Port Select 0 and Port Select 1. These input signals, in conjunction with the RD and WR inputs, control the selection of one of the three ports or the control word registers. They are normally connected to the least significant bits of the address bus (A_0 and A_1).

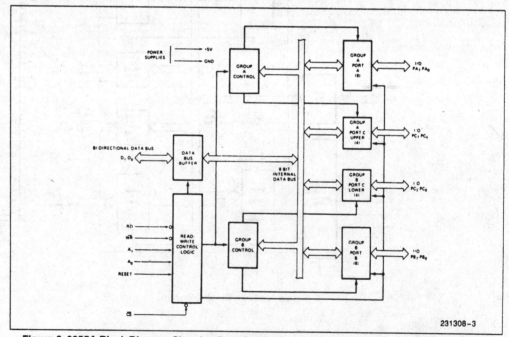

231308-3

Figure 3. 8255A Block Diagram Showing Data Bus Buffer and Read/Write Control Logic Functions

8255A BASIC OPERATION

A₁	A₀	$\overline{RD}$	$\overline{WR}$	$\overline{CS}$	Input Operation (READ)
0	0	0	1	0	Port A → Data Bus
0	1	0	1	0	Port B → Data Bus
1	0	0	1	0	Port C → Data Bus
					Output Operation (WRITE)
0	0	1	0	0	Data Bus → Port A
0	1	1	0	0	Data Bus → Port B
1	0	1	0	0	Data Bus → Port C
1	1	1	0	0	Data Bus → Control
					Disable Function
X	X	X	X	1	Data Bus → 3-State
1	1	0	1	0	Illegal Condition
X	X	1	1	0	Data Bus → 3-State

(RESET)

Reset. A "high" on this input clears the control register and all ports (A, B, C) are set to the input mode.

Group A and Group B Controls

The functional configuration of each port is programmed by the systems software. In essence, the CPU "outputs" a control word to the 8255A. The control word contains information such as "mode", "bit set", "bit reset", etc., that initializes the functional configuration of the 8255A.

Each of the Control blocks (Group A and Group B) accepts "commands" from the Read/Write Control Logic, receives "control words" from the internal data bus and issues the proper commands to its associated ports.

Control Group A—Port A and Port C upper (C7–C4)
Control Group B—Port B and Port C lower (C3–C0)

The Control Word Register can **Only** be written into. No Read operation of the Control Word Register is allowed.

Ports A, B, and C

The 8255A contains three 8-bit ports (A, B, and C). All can be configured in a wide variety of functional characteristics by the system software but each has its own special features or "personality" to further enhance the power and flexibility of the 8255A.

Port A. One 8-bit data output latch/buffer and one 8-bit data input latch.

Port B. One 8-bit data input/output latch/buffer and one 8-bit data input buffer.

Port C. One 8-bit data output latch/buffer and one 8-bit data input buffer (no latch for input). This port can be divided into two 4-bit ports under the mode control. Each 4-bit port contains a 4-bit latch and it can be used for the control signal outputs and status signal inputs in conjunction with ports A and B.

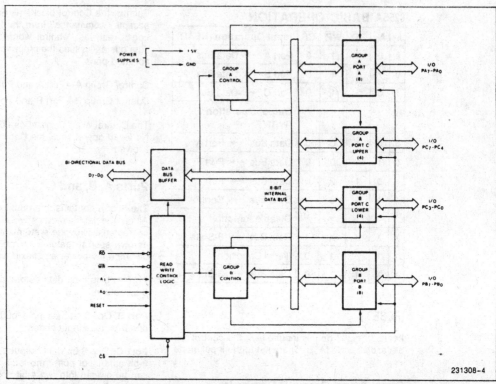

Figure 4. 8225A Block Diagram Showing Group A and Group B Control Functions

Pin Configuration

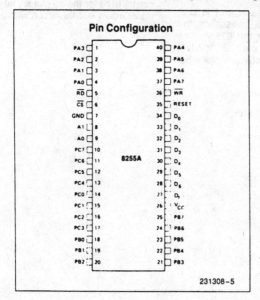

8255A

231308–5

Pin Names

D$_7$–D$_0$	Data Bus (Bi-Directional)
RESET	Reset Input
$\overline{CS}$	Chip Select
$\overline{RD}$	Read Input
$\overline{WR}$	Write Input
A0, A1	Port Address
PA7–PA0	Port A (BIT)
PB7–PB0	Port B (BIT)
PC7–PC0	Port C (BIT)
V$_{CC}$	+ 5 Volts
GND	0 Volts

8255A OPERATIONAL DESCRIPTION

Mode Selection

There are three basic modes of operation that can be selected by the system software:

Mode 0—Basic Input/Output

Mode 1—Strobed Input/Output

Mode 2—Bi-Directional Bus

When the reset input goes "high" all ports will be set to the input mode (i.e., all 24 lines will be in the high impedance state). After the reset is removed the 8255A can remain in the input mode with no additional initialization required. During the execution of the system program any of the other modes may be selected using a single output instruction. This allows a single 8255A to service a variety of peripheral devices with a simple software maintenance routine.

The modes for Port A and Port B can be separately defined, while Port C is divided into two portions as required by the Port A and Port B definitions. All of the output registers, including the status flip-flops, will be reset whenever the mode is changed. Modes may be combined so that their functional definition can be "tailored" to almost any I/O structure. For instance; Group B can be programmed in Mode 0 to monitor simple switch closings or display computational results, Group A could be programmed in Mode 1 to monitor a keyboard or tape reader on an interrupt-driven basis.

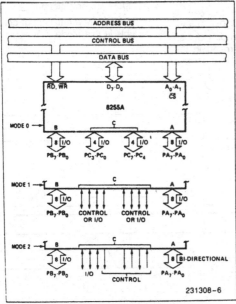

Figure 5. Basic Mode Definitions and Bus Interface

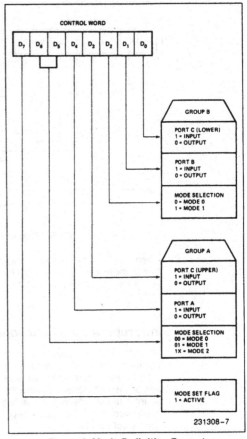

Figure 6. Mode Definition Format

The mode definitions and possible mode combinations may seem confusing at first but after a cursory review of the complete device operation a simple, logical I/O approach will surface. The design of the 8255A has taken into account things such as efficient PC board layout, control signal definition vs PC layout and complete functional flexibility to support almost any peripheral device with no external logic. Such design represents the maximum use of the available pins.

Single Bit Set/Reset Feature

Any of the eight bits of Port C can be Set or Reset using a single OUTput instruction. This feature reduces software requirements in Control-based applications.

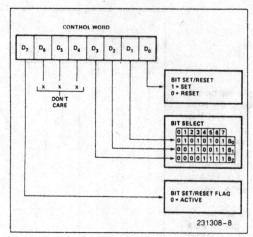

Figure 7. Bit Set/Reset Format

When Port C is being used as status/control for Port A or B, these bits can be set or reset by using the Bit Set/Reset operation just as if they were data output ports.

Interrupt Control Functions

When the 8255A is programmed to operate in mode 1 or mode 2, control signals are provided that can be used as interrupt request inputs to the CPU. The interrupt request signals, generated from port C, can be inhibited or enabled by setting or resetting the associated INTE flip-flop, using the bit set/reset function of port C.

This function allows the Programmer to disallow or allow a specific I/O device to interrupt the CPU without affecting any other device in the interrupt structure.

INTE flip-flop definition:

(BIT-SET)—INTE is set—Interrupt enable

(BIT-RESET)—INTE is RESET—Interrupt disable

NOTE:
All Mask flip-flops are automatically reset during mode selection and device Reset.

Operating Modes

MODE 0 (Basic Input/Output). This functional configuration provides simple input and output operations for each of the three ports. No "handshaking" is required, data is simply written to or read from a specified port.

Mode 0 Basic Functional Definitions:

- Two 8-bit ports and two 4-bit ports.
- Any port can be input or output.
- Outputs are latched.
- Inputs are not latched.
- 16 different Input/Output configurations are possible in this Mode.

MODE 0 (BASIC INPUT)

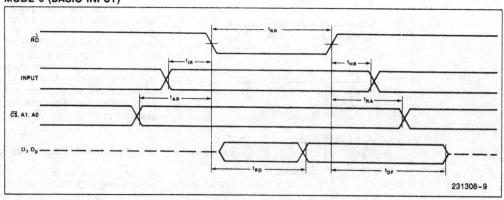

MODE 0 (BASIC OUTPUT)

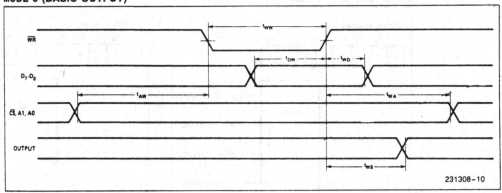

231308-10

MODE 0 PORT DEFINITION

A		B		Group A			Group B	
D_4	D_3	D_1	D_0	Port A	Port C (Upper)	#	Port B	Port C (Lower)
0	0	0	0	OUTPUT	OUTPUT	0	OUTPUT	OUTPUT
0	0	0	1	OUTPUT	OUTPUT	1	OUTPUT	INPUT
0	0	1	0	OUTPUT	OUTPUT	2	INPUT	OUTPUT
0	0	1	1	OUTPUT	OUTPUT	3	INPUT	INPUT
0	1	0	0	OUTPUT	INPUT	4	OUTPUT	OUTPUT
0	1	0	1	OUTPUT	INPUT	5	OUTPUT	INPUT
0	1	1	0	OUTPUT	INPUT	6	INPUT	OUTPUT
0	1	1	1	OUTPUT	INPUT	7	INPUT	INPUT
1	0	0	0	INPUT	OUTPUT	8	OUTPUT	OUTPUT
1	0	0	1	INPUT	OUTPUT	9	OUTPUT	INPUT
1	0	1	0	INPUT	OUTPUT	10	INPUT	OUTPUT
1	0	1	1	INPUT	OUTPUT	11	INPUT	INPUT
1	1	0	0	INPUT	INPUT	12	OUTPUT	OUTPUT
1	1	0	1	INPUT	INPUT	13	OUTPUT	INPUT
1	1	1	0	INPUT	INPUT	14	INPUT	OUTPUT
1	1	1	1	INPUT	INPUT	15	INPUT	INPUT

MODE CONFIGURATIONS

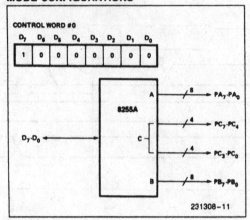

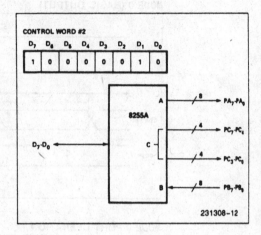

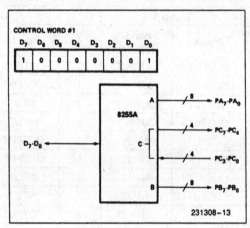

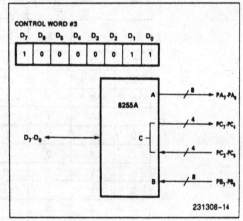

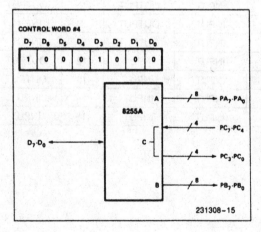

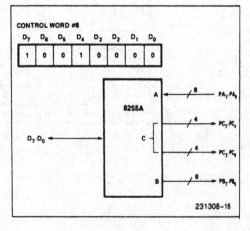

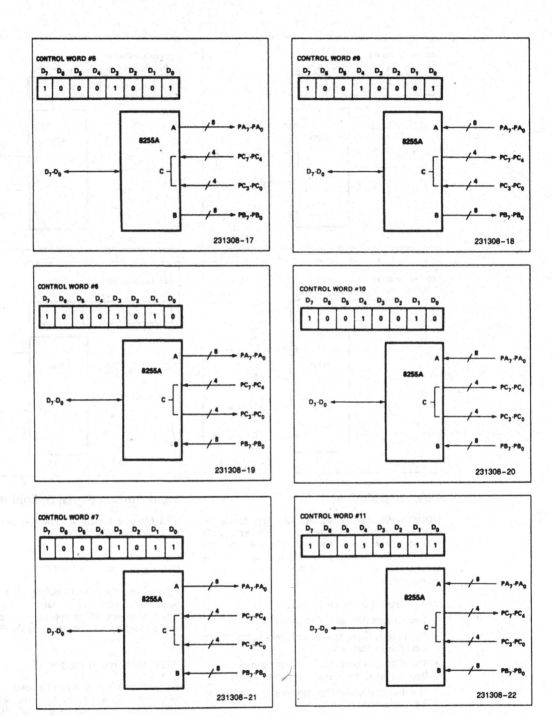

CONTROL WORD #8

D_7	D_6	D_5	D_4	D_3	D_2	D_1	D_0
1	0	0	0	1	0	0	1

231308-17

CONTROL WORD #9

D_7	D_6	D_5	D_4	D_3	D_2	D_1	D_0
1	0	0	1	0	0	0	1

231308-18

CONTROL WORD #6

D_7	D_6	D_5	D_4	D_3	D_2	D_1	D_0
1	0	0	0	1	0	1	0

231308-19

CONTROL WORD #10

D_7	D_6	D_5	D_4	D_3	D_2	D_1	D_0
1	0	0	1	0	0	1	0

231308-20

CONTROL WORD #7

D_7	D_6	D_5	D_4	D_3	D_2	D_1	D_0
1	0	0	0	1	0	1	1

231308-21

CONTROL WORD #11

D_7	D_6	D_5	D_4	D_3	D_2	D_1	D_0
1	0	0	1	0	0	1	1

231308-22

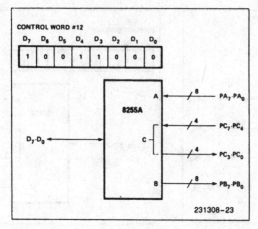

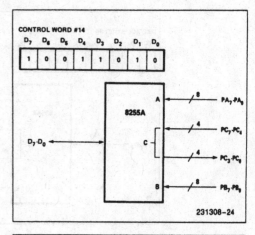

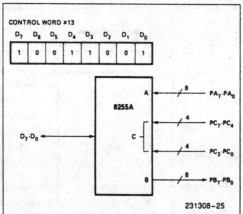

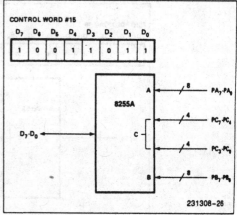

Operating Modes

MODE 1 (Strobed Input/Output). This functional configuration provides a means for transferring I/O data to or from a specified port in conjunction with strobes or "handshaking" signals. In mode 1, port A and port B use the lines on port C to generate or accept these "handshaking" signals.

Mode 1 Basic Functional Definitions:

- Two Groups (Group A and Group B)
- Each group contains one 8-bit data port and one 4-bit control/data port.
- The 8-bit data port can be either input or output. Both inputs and outputs are latched.
- The 4-bit port is used for control and status of the 8-bit data port.

Input Control Signal Definition

$\overline{STB}$ **(Strobe Input).** A "low" on this input loads data into the input latch.

IBF (Input Buffer Full F/F)

A "high" on this output indicates that the data has been loaded into the input latch; in essence, an acknowledgement. IBF is set by STB input being low and is reset by the rising edge of the RD input.

INTR (Interrupt Request)

A "high" on this output can be used to interrupt the CPU when an input device is requesting service. INTR is set by the $\overline{STB}$ is a "one", IBF is a "one" and INTE is a "one". It is reset by the falling edge of RD. This procedure allows an input device to request service from the CPU by simply strobing its data into the port.

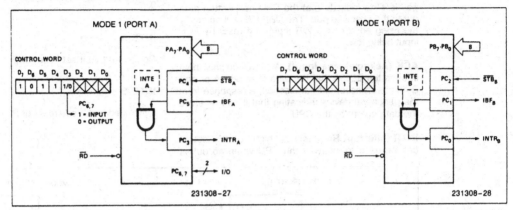

Figure 8. MODE 1 Input

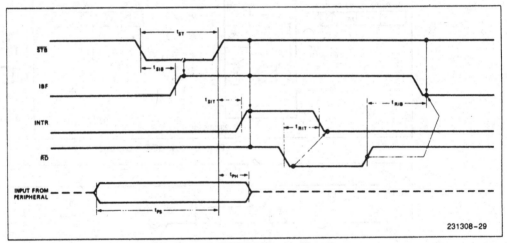

Figure 9. MODE 1 (Strobed Input)

Output Control Signal Definition

$\overline{OBF}$ **(Output Buffer Full F/F).** The $\overline{OBF}$ output will go "low" to indicate that the CPU has written data out to the specified port. The $\overline{OBF}$ F/F will be set by the rising edge of the $\overline{WR}$ input and reset by $\overline{ACK}$ input being low.

$\overline{ACK}$ **(Acknowledge Input).** A "low" on this input informs the 8255A that the data from port A or port B has been accepted. In essence, a response from the peripheral device indicating that it has received the data output by the CPU.

INTR (Interrupt Request). A "high" on this output can be used to interrupt the CPU when an output device has accepted data transmitted by the CPU. INTR is set when $\overline{ACK}$ is a "one", $\overline{OBF}$ is a "one", and INTE is a "one". It is reset by the falling edge of $\overline{WR}$.

INTE A

Controlled by bit set/reset of PC_6.

INTE B

Controlled by bit set/reset of PC_2.

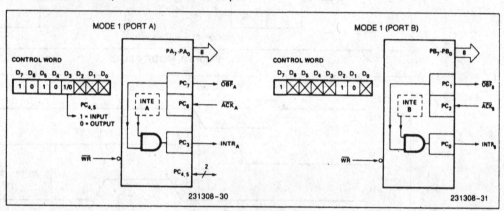

Figure 10. MODE 1 Output

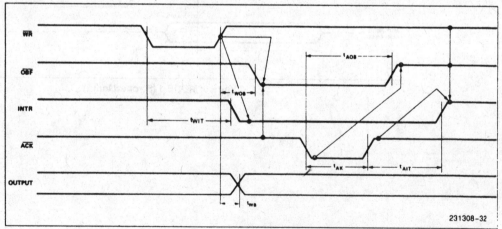

Figure 11. MODE 1 (Strobed Output)

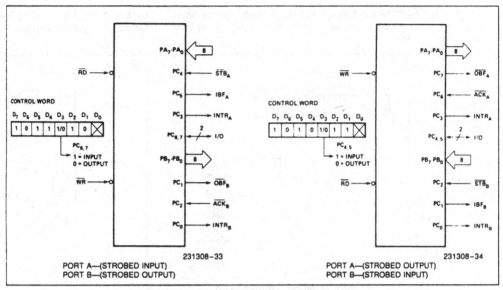

Figure 12. Combinations of MODE 1

Combinations of MODE 1

Port A and Port B can be individually defined as input or output in MODE 1 to support a wide variety of strobed I/O applications.

Operating Modes

MODE 2 (Strobed Bidirectional Bus I/O). This functional configuration provides a means for communicating with a peripheral device or structure on a single 8-bit bus for both transmitting and receiving data (bidirectional bus I/O). "Handshaking" signals are provided to maintain proper bus flow discipline in a similar manner to MODE 1. Interrupt generation and enable/disable functions are also available.

MODE 2 Basic Functional Definitions:

- Used in Group A **only**.
- One 8-bit, bi-directional bus Port (Port A) and a 5-bit control Port (Port C).
- Both inputs and outputs are latched.
- The 5-bit control port (Port C) is used for control and status for the 8-bit, bi-directional bus port (Port A).

Bidirectional Bus I/O Control Signal Definition

INTR (Interrupt Request). A high on this output can be used to interrupt the CPU for both input or output operations.

Output Operations

OBF (Output Buffer Full). The $\overline{OBF}$ output will go "low" to indicate that the CPU has written data out to port A.

ACK (Acknowledge). A "low" on this input enables the tri-state output buffer of port A to send out the data. Otherwise, the output buffer will be in the high impedance state.

INTE 1 (The INTE Flip-Flop Associated with OBF). Controlled by bit set/reset of PC_6.

Input Operations

STB (Strobe Input). A "low" on this input loads data into the input latch.

IBF (Input Buffer Full F/F). A "high" on this output indicates that data has been loaded into the input latch.

INTE 2 (The INTE Flip-Flop Associated with IBF). Controlled by bit set/reset of PC_4.

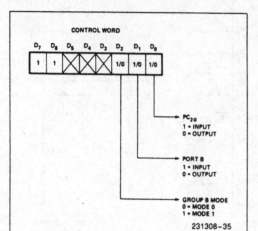

Figure 13. MODE Control Word

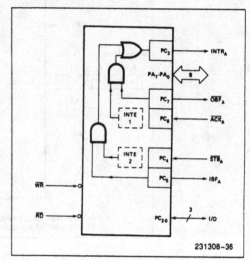

Figure 14. MODE 2

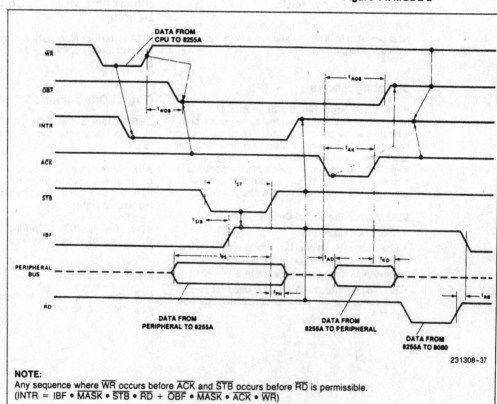

NOTE:
Any sequence where $\overline{WR}$ occurs before $\overline{ACK}$ and $\overline{STB}$ occurs before $\overline{RD}$ is permissible.
(INTR = IBF • $\overline{MASK}$ • $\overline{STB}$ • $\overline{RD}$ + $\overline{OBF}$ • $\overline{MASK}$ • $\overline{ACK}$ • $\overline{WR}$)

Figure 15. MODE 2 (Bidirectional)

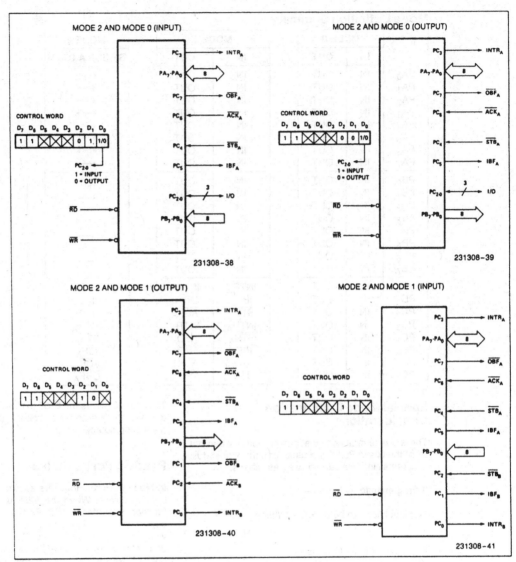

Figure 16. MODE ¼ Combinations

Mode Definition Summary

	MODE 0		MODE 1		MODE 2
	IN	OUT	IN	OUT	GROUP A ONLY
PA_0	IN	OUT	IN	OUT	↔
PA_1	IN	OUT	IN	OUT	↔
PA_2	IN	OUT	IN	OUT	↔
PA_3	IN	OUT	IN	OUT	↔
PA_4	IN	OUT	IN	OUT	↔
PA_5	IN	OUT	IN	OUT	↔
PA_6	IN	OUT	IN	OUT	↔
PA_7	IN	OUT	IN	OUT	↔
PB_0	IN	OUT	IN	OUT	—
PB_1	IN	OUT	IN	OUT	—
PB_2	IN	OUT	IN	OUT	—
PB_3	IN	OUT	IN	OUT	—
PB_4	IN	OUT	IN	OUT	—
PB_5	IN	OUT	IN	OUT	—
PB_6	IN	OUT	IN	OUT	—
PB_7	IN	OUT	IN	OUT	—
PC_0	IN	OUT	$INTR_B$	$INTR_B$	I/O
PC_1	IN	OUT	IBF_B	$\overline{OBF}_B$	I/O
PC_2	IN	OUT	$\overline{STB}_B$	$\overline{ACK}_B$	I/O
PC_3	IN	OUT	$INTR_A$	$INTR_A$	$INTR_A$
PC_4	IN	OUT	$\overline{STB}_A$	I/O	$\overline{STB}_A$
PC_5	IN	OUT	IBF_A	I/O	IBF_A
PC_6	IN	OUT	I/O	$\overline{ACK}_A$	$\overline{ACK}_A$
PC_7	IN	OUT	I/O	$\overline{OBF}_A$	$\overline{OBF}_A$

MODE 0
OR MODE 1
ONLY
(bracket spanning PB rows, MODE 2 column)

Special Mode Combination Considerations

There are several combinations of modes when not all of the bits in Port C are used for control or status. The remaining bits can be used as follows:

If Programmed as Inputs—

All input lines can be accessed during a normal Port C read.

If Programmed as Outputs—

Bits in C upper (PC_7–PC_4) must be individually accessed using the bit set/reset function.

Bits in C lower (PC_3–PC_0) can be accessed using the bit set/reset function or accessed as a threesome by writing into Port C.

Source Current Capability on Port B and Port C

Any set of **eight** output buffers, selected randomly from Ports B and C can source 1 mA at 1.5 volts.

This feature allows the 8255 to directly drive Darlington type drivers and high-voltage displays that require such source current.

Reading Port C Status

In Mode 0, Port C transfers data to or from the peripheral device. When the 8255 is programmed to function in Modes 1 or 2, Port C generates or accepts "hand-shaking" signals with the peripheral device. Reading the contents of Port C allows the programmer to test or verify the "status" of each peripheral device and change the program flow accordingly.

There is no special instruction to read the status information from Port C. A normal read operation of Port C is executed to perform this function.

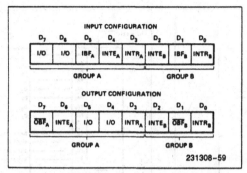

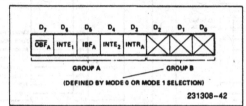

Figure 17. MODE 1 Status Word Format

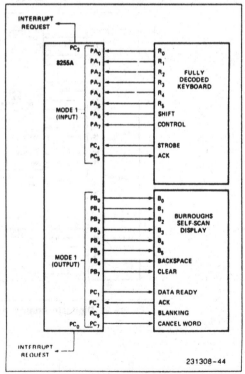

Figure 18. MODE 2 Status Word Format

APPLICATIONS OF THE 8255A

The 8255A is a very powerful tool for interfacing peripheral equipment to the microcomputer system. It represents the optimum use of available pins and is flexible enough to interface almost any I/O device without the need for additional external logic.

Each peripheral device in a microcomputer system usually has a "service routine" associated with it. The routine manages the software interface between the device and the CPU. The functional definition of the 8255A is programmed by the I/O service routine and becomes an extension of the system software. By examining the I/O devices interface characteristics for both data transfer and timing, and matching this information to the examples and tables in the detailed operational description, a control word can easily be developed to initialize the 8255A to exactly "fit" the application. Figures 19 through 25 represent a few examples of typical applications of the 8255A.

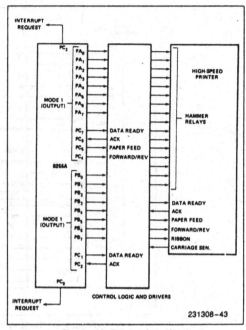

Figure 19. Printer Interface

Figure 20. Keyboard and Display Interface

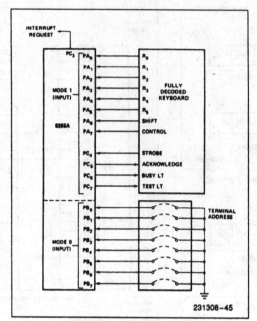

Figure 21. Keyboard and Terminal Address Interface

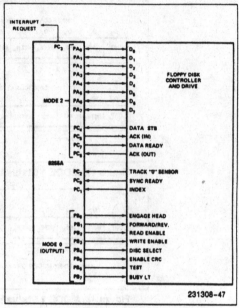

Figure 23. Basic Floppy Disk Interface

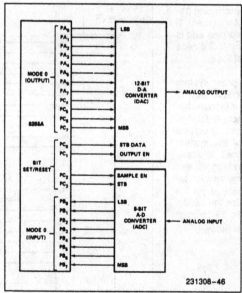

Figure 22. Digital to Analog, Analog to Digital

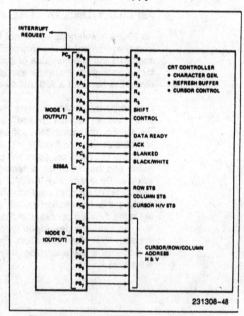

Figure 24. Basic CRT Controller Interface

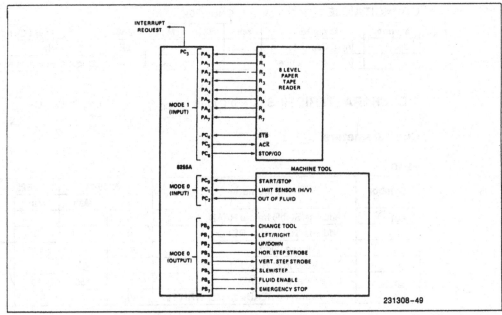

Figure 25. Machine Tool Controller Interface

ABSOLUTE MAXIMUM RATINGS*

Ambient Temperature Under Bias ......0°C to 70°C

Storage Temperature .... −65°C to +150°C

Voltage on Any Pin
  with Respect to Ground.......... −0.5V to +7V

Power Dissipation ....1 Watt

*Notice: Stresses above those listed under "Absolute Maximum Ratings" may cause permanent damage to the device. This is a stress rating only and functional operation of the device at these or any other conditions above those indicated in the operational sections of this specification is not implied. Exposure to absolute maximum rating conditions for extended periods may affect device reliability.

D.C. CHARACTERISTICS T_A = 0°C to 70°C, V_{CC} = +5V ±10%, GND = 0V*

Symbol	Parameter	Min	Max	Unit	Test Conditions
V_{IL}	Input Low Voltage	−0.5	0.8	V	
V_{IH}	Input High Voltage	2.0	·V_{CC}	V	
V_{OL} (DB)	Output Low Voltage (Data Bus)		0.45*	V	I_{OL} = 2.5 mA
V_{OL} (PER)	Output Low Voltage (Peripheral Port)		0.45*	V	I_{OL} = 1.7 mA
V_{OH} (DB)	Output High Voltage (Data Bus)	2.4		V	I_{OH} = −400 μA
V_{OH} (PER)	Output High Voltage (Peripheral Port)	2.4		V	I_{OH} = −200 μA
I_{DAR}[1]	Darlington Drive Current	−1.0	−4.0	mA	R_{EXT} = 750Ω; V_{EXT} = 1.5V
I_{CC}	Power Supply Current		120	mA	
I_{IL}	Input Load Current		±10	μA	V_{IN} = V_{CC} to 0V
I_{OFL}	Output Float Leakage		±10	μA	V_{OUT} = V_{CC} to 0.45V

NOTE:
1. Available on any 8 pins from Port B and C.

CAPACITANCE $T_A = 25°C$, $V_{CC} = GND = 0V$

Symbol	Parameter	Min	Typ	Max	Unit	Test Conditions
C_{IN}	Input Capacitance			10	pF	$f_c = 1$ MHz[4]
$C_{I/O}$	I/O Capacitance			20	pF	Unmeasured pins returned to GND[4]

A.C. CHARACTERISTICS $T_A = 0°C$ to $70°C$, $V_{CC} = +5V \pm 10\%$, $GND = 0V$*

Bus Parameters

READ

Symbol	Parameter	8255A		8255A-5		Unit
		Min	Max	Min	Max	
t_{AR}	Address Stable before READ	0		0		ns
t_{RA}	Address Stable after READ	0		0		ns
t_{RR}	READ Pulse Width	300		300		ns
t_{RD}	Data Valid from READ[1]		250		200	ns
t_{DF}	Data Float after READ	10	150	10	100	ns
t_{RV}	Time between READs and/or WRITEs	850		850		ns

WRITE

Symbol	Parameter	8255A		8255A-5		Unit
		Min	Max	Min	Max	
t_{AW}	Address Stable before WRITE	0		0		ns
t_{WA}	Address Stable after WRITE	20		20		ns
t_{WW}	WRITE Pulse Width	400		300		ns
t_{DW}	Data Valid to WRITE (T.E.)	100		100		ns
t_{WD}	Data Valid after WRITE	30		30		ns

OTHER TIMINGS

Symbol	Parameter	8255A		8255A-5		Unit
		Min	Max	Min	Max	
t_{WB}	WR = 1 to Output[1]		350		350	ns
t_{IR}	Peripheral Data before RD	0		0		ns
t_{HR}	Peripheral Data after RD	0		0		ns
t_{AK}	ACK Pulse Width	300		300		ns
t_{ST}	STB Pulse Width	500		500		ns
t_{PS}	Per. Data before T.E. of STB	0		0		ns
t_{PH}	Per. Data after T.E. of STB	180		180		ns
t_{AD}	ACK = 0 to Output[1]		300		300	ns
t_{KD}	ACK = 1 to Output Float	20	250	20	250	ns

A.C. CHARACTERISTICS (Continued)

OTHER TIMINGS (Continued)

Symbol	Parameter	8255A		8255A-5		Unit
		Min	Max	Min	Max	
t_{WOB}	WR = 1 to OBF = 0[1]		650		650	ns
t_{AOB}	ACK = 0 to OBF = 1[1]		350		350	ns
t_{SIB}	STB = 0 to IBF = 1[1]		300		300	ns
t_{RIB}	RD = 1 to IBF = 0[1]		300		300	ns
t_{RIT}	RD = 0 to INTR = 0[1]		400		400	ns
t_{SIT}	STB = 1 to INTR = 1[1]		300		300	ns
t_{AIT}	ACK = 1 to INTR = 1[1]		350		350	ns
t_{WIT}	WR = 0 to INTR = 0[1, 3]		850		850	ns

NOTES:
1. Test Conditions: C_L = 150 pF.
2. Period of Reset pulse must be at least 50 μs during or after power on. Subsequent Reset pulse can be 500 ns min.
3. INTR ↑ may occur as early as $\overline{WR}$ ↓.
4. Sampled, not 100% tested.
*For Extended Temperature EXPRESS, use M8255A electrical parameters.

A.C. TESTING INPUT, OUTPUT WAVEFORM

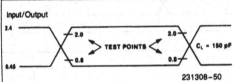

231308-50

A.C. Testing: Inputs are driven at 2.4V for a Logic "1" and 0.45V for a Logic "0". Timing measurements are made at 2.0V for a Logic "1" and 0.8V for a Logic "0".

A.C. TESTING LOAD CIRCUIT

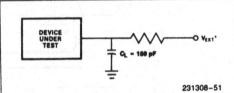

231308-51

*V_{EXT} is set at various voltages during testing to guarantee the specification. C_L includes jig capacitance.

WAVEFORMS

MODE 0 (BASIC INPUT)

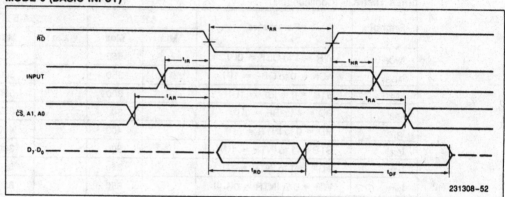

231308-52

MODE 0 (BASIC OUTPUT)

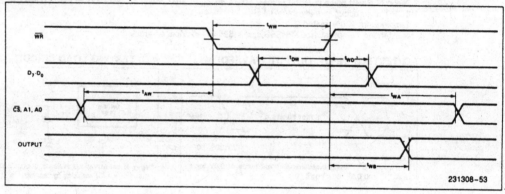

231308-53

WAVEFORMS (Continued)

MODE 1 (STROBED INPUT)

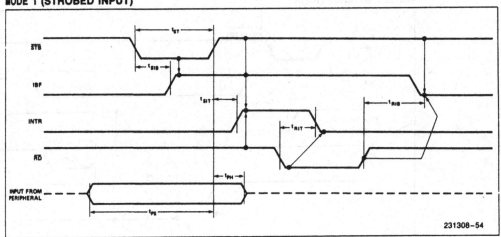

231308-54

MODE 1 (STROBED OUTPUT)

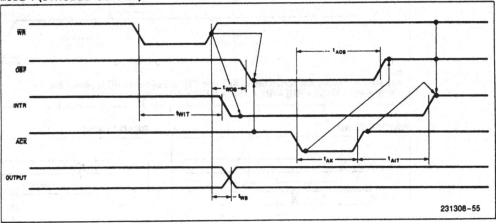

231308-55

WAVEFORMS (Continued)

MODE 2 (BIDIRECTIONAL)

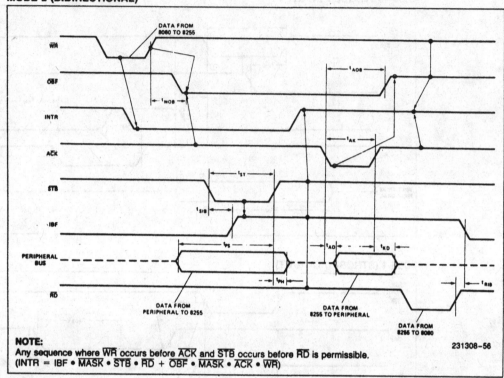

NOTE:
Any sequence where $\overline{WR}$ occurs before $\overline{ACK}$ and $\overline{STB}$ occurs before $\overline{RD}$ is permissible.
(INTR = IBF • $\overline{MASK}$ • $\overline{STB}$ • $\overline{RD}$ + $\overline{OBF}$ • $\overline{MASK}$ • $\overline{ACK}$ • $\overline{WR}$)

231308-56

WRITE TIMING

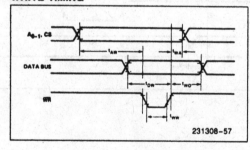

231308-57

READ TIMING

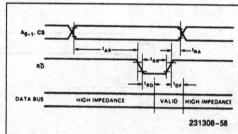

231308-58

Appendix F
Solution for heat flow in one dimension

The problem at hand is to solve the differential equation for heat flow in one dimension, *vis*

$$\partial T/\partial t = \alpha^2(\partial^2 T/\partial z^2) \tag{F.1}$$

where $\alpha^2 = k/s$ and where the rod extends to infinity on both sides. The initial condition is that the temperature at $t = 0$ is given, ie, $T(z, t = 0) = f(z)$ where $f(z)$ is the given initial temperature distribution along the bar.

To proceed, we try the method of separation of variables by writing $T(z, t) = F(z)G(t)$. Equation (F.1) then becomes

$$\frac{\partial G/\partial t}{\alpha^2 G} = \frac{\partial^2 F/\partial z^2}{F} \tag{F.2}$$

Since the variables t and z vary independently, each side of Equation (F.2) must be equal to a constant, say q, giving two ordinary differential equations.

$$\left. \begin{array}{l} dG/dt = q\alpha^2 G \\ d^2F/dz^2 = qF \end{array} \right\} \tag{F.3}$$

The solution for the first is

$$G(t) = K \exp(q\alpha^2 t) \tag{F.4}$$

where K is a constant. If q is positive, this solution grows without limit and thus is not a physically realizable solution. So $q \leq 0$ and we can write it as $q = -p^2$ to force this condition. Equation (F.4) becomes

$$G(t) = K \exp(-p^2\alpha^2 t) \tag{F.5}$$

The second of equations (F.3) can now be recognized as a simple harmonic motion equation

$$(d^2F/dz^2) + p^2F = 0 \tag{F.6}$$

with the solution

$$F(z) = A \cos(pz) + B \sin(pz) \tag{F.7}$$

So, the solution to the differential equation has the form

$$\begin{aligned} T(z, t; p) &= FG \\ &= [A \cos(pz) + B \sin(pz)] \exp(-p^2\alpha^2 t) \end{aligned} \tag{F.8}$$

where the constant K has been absorbed into A and B. Equation (F.8) is true for any p and any linear combination of solutions with different p will also be a solution. In particular, a general solution is

$$T(z, t) = \int_0^\infty [A(p) \cos(pz) + B(p) \sin(pz)] \exp(-p^2\alpha^2 t)dp \tag{F.9}$$

Using the initial condition that $T(z, 0) = f(z)$, gives for Equation (F.9)

$$T(z, 0) = \int_0^\infty [A(p) \cos(pz) + B(p) \sin(pz)]dp \tag{F.10}$$

The Fourier integral theorem gives the following expressions for A and B

$$\left.\begin{aligned} A(p) &= (1/\pi) \int_{-\infty}^\infty f(\xi) \cos(p\xi)d\xi \\ B(p) &= (1/\pi) \int_{-\infty}^\infty f(\xi) \sin(p\xi)d\xi \end{aligned}\right\} \tag{F.11}$$

Using these expressions, Equation (F.9) becomes

$$\begin{aligned} T(z, t) &= (1/\pi) \int_0^\infty \left\{ \int_{-\infty}^\infty f(\xi)[\cos(p\xi) \cos(pz) + \sin(p\xi) \sin(p\xi)] \right. \\ &\qquad\qquad \left. \times \exp(-p^2\alpha^2 t)d\xi \right\}dp \\ &= (1/\pi) \int_0^\infty \left\{ \int_{-\infty}^\infty f(\xi) \cos(pz - p\xi) \exp(-p^2\alpha^2 t)d\xi \right\}dp \end{aligned} \tag{F.12}$$

Exchanging the order of integration gives

$$T(z, t) = (1/\pi) \int_{-\infty}^\infty f(\xi)\left\{ \int_0^\infty \cos(pz - p\xi) \exp(-p^2\alpha^2 t)dp \right\}d\xi$$

The inner integral can be found in a table of integrals and is equal to

$$\frac{\pi^{1/2}}{2\alpha t^{1/2}} \exp\left[-\frac{(z - \xi)^2}{4\alpha^2 t} \right]$$

Therefore

$$T(z, t) = \frac{1}{2\alpha(\pi t)^{1/2}} \int_{-\infty}^\infty f(\xi) \exp\left[-\frac{(z - \xi)^2}{4\alpha^2 t} \right]d\xi \tag{F.13}$$

In the physical situation of a very quick impulse of heat given to a rod at $z = 0$, the initial temperature distribution will be (Figure F.1)

$$f(z) = \lim_{\Delta z \to 0} \begin{cases} 0 & z < -\Delta z \\ T_{max} & -\Delta z < z < \Delta z \\ 0 & \Delta z < z \end{cases}$$

Fig. F.1. Initial temperature distribution on the infinite rod.

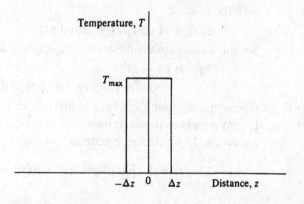

Equation (F.13) becomes

$$T(z, t) = \frac{1}{2\alpha(\pi t)^{1/2}} \int_{-\Delta z}^{\Delta z} f(\xi) \exp\left[-\frac{(z - \xi)^2}{4\alpha^2 t}\right] d\xi \qquad \text{(F.14)}$$

If Δz is small, the exponential in the integral will not vary much across the interval $-\Delta z$ to Δz and so may be evaluated at $\xi = 0$ and be removed from the integral.

$$T(z, t) = \frac{1}{2\alpha(\pi t)^{1/2}} \exp\left(-\frac{z^2}{4\alpha^2 t}\right) \int_{-\Delta z}^{\Delta z} f(\xi) d\xi \qquad \text{(F.15)}$$

The remaining integral is just a constant so

$$T(z, t) = \frac{B}{t^{1/2}} \exp\left(-\frac{z^2}{4\alpha^2 t}\right) \qquad \text{(F.16)}$$

where B has absorbed all the constants. Also any constant value, say A, is a solution to the differential equation, so

$$T(z, t) = A + \frac{B}{t^{1/2}} \exp\left(-\frac{z^2}{4\alpha^2 t}\right) \qquad \text{(F.17)}$$

as is stated as Equation (5.1.5).

That's all folks.

Appendix G
Finite impulse heat flow in a rod

Equation (5.1.9) describes the flow of heat in a rod when the heat is applied very quickly at one point. The term very quickly means that the ratio of the time that the heater is on (call it τ) to the characteristic time of the system, t_1, is much less than one.

$$\tau/t_1 \ll 1 \tag{G.1}$$

Physically, this means that the heat was put into the rod much faster than it flowed away from the point where it was added.

In doing the experiment, equation (G.1) does not always strictly hold. An impulse of 0.5 s gives a τ/t_1 of about 0.4. In that case, the input of heat can be considered to be made up of a series of heat impulses, each of which has a width $\Delta\tau$ such that

$$\Delta\tau/t_1 \ll 1$$

See Figure G.1.

Thus for each of these smaller intervals $\Delta\tau$, Equation (5.1.9) will hold but must be rewritten with a change of origin:

$$T_i = T_1^i \left(\frac{t_1}{t + \tau}\right)^{1/2} \exp\left(\frac{-t_1}{t + \tau}\right) \tag{G.2}$$

where

$$T_1^i = 2q/Azs\pi^{1/2}$$

and $q = P\Delta\tau$ is the heat put in during one interval and P is the power (assumed to be constant). The total temperature change will be given by the sum of the individual T_i:

$$T = \sum_i T_i$$

Fig. G.1. Heat input pulse with finite duration.

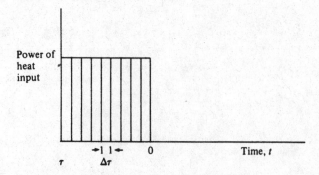

Power of heat input — $\rightarrow$| 1 |$\leftarrow$ 0   Time, t
τ $\Delta\tau$

and the total heat input is

$$Q = \sum_i q$$

If $\Delta\tau$ goes to 0 then the sum goes to an integral:

$$T(t) = \int_0^T T_1' \left(\frac{t_1}{t+\tau}\right)^{1/2} \exp\left(\frac{-t_1}{t+\tau}\right) d\tau \tag{G.3}$$

with $T_1' = 2P/Azs\pi^{1/2}$.

By a suitable change in variable and integration by parts, this integral can be evaluated giving

$$T = 2T_1 \frac{1}{\gamma} \left[(x+\gamma)^{1/2} \exp\left(\frac{-1}{x+\gamma}\right) - x^{1/2} \exp\left(\frac{-1}{x}\right) \right.$$
$$\left. + \pi^{1/2} \, \text{erf}\left(\frac{1}{x+\gamma}\right)^{1/2} - \pi^{1/2} \, \text{erf}\left(\frac{1}{x}\right) \right] \tag{G.4}$$

where

$$T_1 = \frac{2Q}{Azs\pi^{1/2}}$$

as in Equation (5.1.9)

$$\gamma = \tau/t_1$$
$$x = t/t_1$$

and

$$\text{erf}(\eta) = \frac{2}{\pi^{1/2}} \int_0^\eta \exp(-\xi^2) d\xi$$

is the error function which can be evaluated using a table or a computer program.

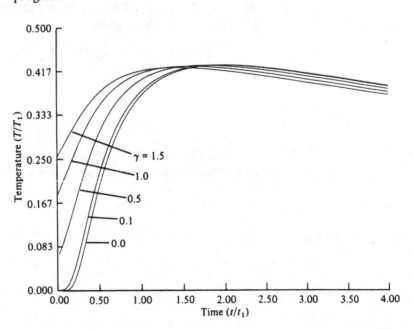

Fig. G.2. Heat flow for a finite heat pulse of length $\gamma = \tau/t_1$ with $t = 0$ at the end of the pulse.

Fig. G.3. As in Figure G.2 but with $t = 0$ in the middle of the pulse.

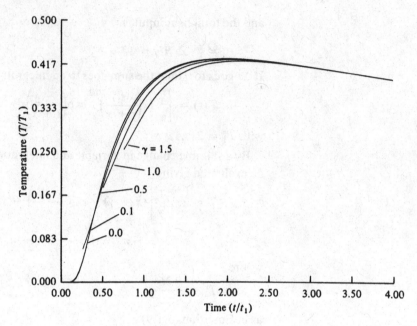

Figure G.2 is a plot of T/T_1 vs. t/t_1 for $\gamma = 0.01$–1.5 and shows the error which is made when modeling an experiment with the impulse solution (Equation (5.1.9)) when Equation (G.4) is actually more correct. The curve with $\gamma = 0.01$ is essentially equal to the impulse solution Equation (5.1.9). For ratios of $\gamma > 0.1$ an appreciable error is made.

If $t = 0$ is measured from the center of the finite input pulse, a better fit is obtained. Equation (G.4) can be translated to this new origin by the substitution $t \rightarrow t - \tau/2$ giving

$$T = 2T_1 \frac{1}{\gamma} \left[(x + \gamma/2)^{1/2} \exp\left(\frac{-1}{x + \gamma/2}\right) \right.$$
$$- (x - \gamma/2)^{1/2} \exp\left(\frac{-1}{x - \gamma/2}\right)$$
$$\left. + \pi^{1/2} \operatorname{erf}\left(\frac{1}{x + \gamma/2}\right) - \pi^{1/2} \operatorname{erf}\left(\frac{1}{x - \gamma/2}\right) \right] \qquad \text{(G.5)}$$

A plot of Equation (G.5) is given in Figure G.3. Ratios of up to $\gamma = 1$ can be tolerated without appreciable error with this time origin.

Appendix H
8088 Microprocessor data sheets

Here is a description of the 8088/8086 microprocessor instruction set.

Table 2-21. Instruction Set Reference Data

AAA	AAA (no operands) ASCII adjust for addition				Flags	O D I T S Z A P C U U U X U X
Operands		**Clocks**	**Transfers***	**Bytes**	**Coding Example**	
(no operands)		4	—	1	AAA	

AAD	AAD (no operands) ASCII adjust for division				Flags	O D I T S Z A P C U X X U X U
Operands		**Clocks**	**Transfers***	**Bytes**	**Coding Example**	
(no operands)		60	—	2	AAD	

AAM	AAM (no operands) ASCII adjust for multiply				Flags	O D I T S Z A P C U X X U X U
Operands		**Clocks**	**Transfers***	**Bytes**	**Coding Example**	
(no operands)		83	—	1	AAM	

AAS	AAS (no operands) ASCII adjust for subtraction				Flags	O D I T S Z A P C U U U X U X
Operands		**Clocks**	**Transfers***	**Bytes**	**Coding Example**	
(no operands)		4	—	1	AAS	

*For the 8086, add four clocks for each 16-bit word transfer with an odd address. For the 8088, add four clocks for each 16-bit word transfer.

Mnemonics © Intel, 1978

Table 2-21. Instruction Set Reference Data (Cont'd.)

ADC		ADC destination,source Add with carry			Flags	O D I T S Z A P C X X X X X X
Operands		**Clocks**	**Transfers***	**Bytes**		**Coding Example**
register, register		3	—	2		ADC AX, SI
register, memory		9 + EA	1	2-4		ADC DX, BETA [SI]
memory, register		16 + EA	2	2-4		ADC ALPHA [BX] [SI], DI
register, immediate		4	—	3-4		ADC BX, 256
memory, immediate		17 + EA	2	3-6		ADC GAMMA, 30H
accumulator, immediate		4	—	2-3		ADC AL, 5

ADD		ADD destination,source Addition			Flags	O D I T S Z A P C X X X X X X
Operands		**Clocks**	**Transfers***	**Bytes**		**Coding Example**
register, register		3	—	2		ADD CX, DX
register, memory		9 + EA	1	2-4		ADD DI, [BX].ALPHA
memory, register		16 + EA	2	2-4		ADD TEMP, CL
register, immediate		4	—	3-4		ADD CL, 2
memory, immediate		17 + EA	2	3-6		ADD ALPHA, 2
accumulator, immediate		4	—	2-3		ADD AX, 200

AND		AND destination,source Logical and			Flags	O D I T S Z A P C 0 X X U X 0
Operands		**Clocks**	**Transfers***	**Bytes**		**Coding Example**
register, register		3	—	2		AND AL,BL
register, memory		9 + EA	1	2-4		AND CX,FLAG_WORD
memory, register		16 + EA	2	2-4		AND ASCII [DI],AL
register, immediate		4	—	3-4		AND CX,0F0H
memory, immediate		17 + EA	2	3-6		AND BETA, 01H
accumulator, immediate		4	—	2-3		AND AX, 01010000B

CALL		CALL target Call a procedure			Flags	O D I T S Z A P C
Operands		**Clocks**	**Transfers***	**Bytes**		**Coding Examples**
near-proc		19	1	3		CALL NEAR_PROC
far-proc		28	2	5		CALL FAR_PROC
memptr 16		21 + EA	2	2-4		CALL PROC_TABLE [SI]
regptr 16		16	1	2		CALL AX
memptr 32		37 + EA	4	2-4		CALL [BX].TASK [SI]

CBW		CBW (no operands) Convert byte to word			Flags	O D I T S Z A P C
Operands		**Clocks**	**Transfers***	**Bytes**		**Coding Example**
(no operands)		2	—	1		CBW

*For the 8086, add four clocks for each 16-bit word transfer with an odd address For the 8088, add four clocks for each 16-bit word transfer

Table 2-21. Instruction Set Reference Data (Cont'd.)

CLC	CLC (no operands) Clear carry flag				Flags	O D I T S Z A P C 0
Operands		**Clocks**	**Transfers***	**Bytes**	**Coding Example**	
(no operands)		2	—	1	CLC	

CLD	CLD (no operands) Clear direction flag				Flags	O D I T S Z A P C 0
Operands		**Clocks**	**Transfers***	**Bytes**	**Coding Example**	
(no operands)		2	—	1	CLD	

CLI	CLI (no operands) Clear interrupt flag				Flags	O D I T S Z A P C 0
Operands		**Clocks**	**Transfers***	**Bytes**	**Coding Example**	
(no operands)		2	—	1	CLI	

CMC	CMC (no operands) Complement carry flag				Flags	O D I T S Z A P C X
Operands		**Clocks**	**Transfers***	**Bytes**	**Coding Example**	
(no operands)		2	—	1	CMC	

CMP	CMP destination,source Compare destination to source				Flags	O D I T S Z A P C X X X X X X
Operands		**Clocks**	**Transfers***	**Bytes**	**Coding Example**	
register, register		3	—	2	CMP BX, CX	
register, memory		9 + EA	1	2-4	CMP DH, ALPHA	
memory, register		9 + EA	1	2-4	CMP [BP + 2], SI	
register, immediate		4	—	3-4	CMP BL, 02H	
memory, immediate		10 + EA	1	3-6	CMP [BX].RADAR [DI], 3420H	
accumulator, immediate		4	—	2-3	CMP AL, 00010000B	

CMPS	CMPS dest-string,source-string Compare string				Flags	O D I T S Z A P C X X X X X X
Operands		**Clocks**	**Transfers***	**Bytes**	**Coding Example**	
dest-string, source-string		22	2	1	CMPS BUFF1, BUFF2	
(repeat) dest-string, source-string		9 + 22/rep	2/rep	1	REPE CMPS ID, KEY	

*For the 8086, add four clocks for each 16-bit word transfer with an odd address. For the 8088, add four clocks for each 16-bit word transfer.

Table 2-21. Instruction Set Reference Data (Cont'd.)

CWD

CWD (no operands) Convert word to doubleword				Flags O D I T S Z A P C	
Operands	**Clocks**	**Transfers***	**Bytes**	**Coding Example**	
(no operands)	5	—	1	CWD	

DAA

DAA (no operands) Decimal adjust for addition				Flags O D I T S Z A P C X X X X X X	
Operands	**Clocks**	**Transfers***	**Bytes**	**Coding Example**	
(no operands)	4	—	1	DAA	

DAS

DAS (no operands) Decimal adjust for subtraction				Flags O D I T S Z A P C U X X X X X	
Operands	**Clocks**	**Transfers***	**Bytes**	**Coding Example**	
(no operands)	4	—	1	DAS	

DEC

DEC destination Decrement by 1				Flags O D I T S Z A P C X X X X X	
Operands	**Clocks**	**Transfers***	**Bytes**	**Coding Example**	
reg16	2	—	1	DEC AX	
reg8	3	—	2	DEC AL	
memory	15 + EA	2	2-4	DEC ARRAY [SI]	

DIV

DIV source Division, unsigned				Flags O D I T S Z A P C U U U U U U	
Operands	**Clocks**	**Transfers***	**Bytes**	**Coding Example**	
reg8	80-90	—	2	DIV CL	
reg16	144-162	—	2	DIV BX	
mem8	(86-96) + EA	1	2-4	DIV ALPHA	
mem16	(150-168) + EA	1	2-4	DIV TABLE [SI]	

ESC

ESC external-opcode,source Escape				Flags O D I T S Z A P C	
Operands	**Clocks**	**Transfers***	**Bytes**	**Coding Example**	
immediate, memory	8 + EA	1	2-4	ESC 6,ARRAY [SI]	
immediate, register	2	—	2	ESC 20,AL	

*For the 8086, add four clocks for each 16-bit word transfer with an odd address. For the 8088, add four clocks for each 16-bit word transfer.

Table 2-21. Instruction Set Reference Data (Cont'd.)

HLT	HLT (no operands) Halt			Flags	O D I T S Z A P C
Operands		Clocks	Transfers*	Bytes	Coding Example
(no operands)		2	—	1	HLT

IDIV	IDIV source Integer division			Flags	O D I T S Z A P C U U U U U
Operands		Clocks	Transfers*	Bytes	Coding Example
reg8		101-112	—	2	IDIV BL
reg16		165-184	—	2	IDIV CX
mem8		(107-118) + EA	1	2-4	IDIV DIVISOR__BYTE [SI]
mem16		(171-190) + EA	1	2-4	IDIV [BX].DIVISOR__WORD

IMUL	IMUL source Integer multiplication			Flags	O D I T S Z A P C X U U U U X
Operands		Clocks	Transfers*	Bytes	Coding Example
reg8		80-98	—	2	IMUL CL
reg16		128-154	—	2	IMUL BX
mem8		(86-104) + EA	1	2-4	IMUL RATE__BYTE
mem16		(134-160) + EA	1	2-4	IMUL RATE__WORD [BP] [DI]

IN	IN accumulator,port Input byte or word			Flags	O D I T S Z A P C
Operands		Clocks	Transfers*	Bytes	Coding Example
accumulator, immed8		10	1	2	IN AL, 0FFEAH
accumulator, DX		8	1	1	IN AX, DX

INC	INC destination Increment by 1			Flags	O D I T S Z A P C X X X X X
Operands		Clocks	Transfers*	Bytes	Coding Example
reg16		2	—	1	INC CX
reg8		3	—	2	INC BL
memory		15 + EA	2	2-4	INC ALPHA [DI] [BX]

*For the 8086, add four clocks for each 16-bit word transfer with an odd address. For the 8088, add four clocks for each 16-bit word transfer.

Table 2-21. Instruction Set Reference Data (Cont'd.)

INT	INT interrupt-type Interrupt			Flags	O D I T S Z A P C 0 0
Operands		Clocks	Transfers*	Bytes	Coding Example
immed8 (type = 3)		52	5	1	INT 3
immed8 (type ≠ 3)		51	5	2	INT 67

INTR †	INTR (external maskable interrupt) Interrupt if INTR and IF=1			Flags	O D I T S Z A P C 0 0
Operands		Clocks	Transfers*	Bytes	Coding Example
(no operands)		61	7	N/A	N/A

INTO	INTO (no operands) Interrupt if overflow			Flags	O D I T S Z A P C 0 0
Operands		Clocks	Transfers*	Bytes	Coding Example
(no operands)		53 or 4	5	1	INTO

IRET	IRET (no operands) Interrupt Return			Flags	O D I T S Z A P C R R R R R R R R
Operands		Clocks	Transfers*	Bytes	Coding Example
(no operands)		24	3	1	IRET

JA/JNBE	JA/JNBE short-label Jump if above/Jump if not below nor equal			Flags	O D I T S Z A P C
Operands		Clocks	Transfers*	Bytes	Coding Example
short-label		16 or 4	—	2	JA ABOVE

JAE/JNB	JAE/JNB short-label Jump if above or equal/Jump if not below			Flags	O D I T S Z A P C
Operands		Clocks	Transfers*	Bytes	Coding Example
short-label		16 or 4	—	2	JAE ABOVE__EQUAL

JB/JNAE	JB/JNAE short-label Jump if below/Jump if not above nor equal			Flags	O D I T S Z A P C
Operands		Clocks	Transfers*	Bytes	Coding Example
short-label		16 or 4	—	2	JB BELOW

*For the 8086, add four clocks for each 16-bit word transfer with an odd address. For the 8088, add four clocks for each 16-bit word transfer.

†INTR is not an instruction; it is included in table 2-21 only for timing information.

Table 2-21. Instruction Set Reference Data (Cont'd.)

JBE/JNA		JBE/JNA short-label Jump if below or equal/Jump if not above			Flags	O D I T S Z A P C
Operands		Clocks	Transfers*	Bytes	Coding Example	
short-label		16 or 4	—	2	JNA NOT__ABOVE	

JC		JC short-label Jump if carry			Flags	O D I T S Z A P C
Operands		Clocks	Transfers*	Bytes	Coding Example	
short-label		16 or 4	—	2	JC CARRY__SET	

JCXZ		JCXZ short-label Jump if CX is zero			Flags	O D I T S Z A P C
Operands		Clocks	Transfers*	Bytes	Coding Example	
short-label		18 or 6	—	2	JCXZ COUNT__DONE	

JE/JZ		JE/JZ short-label Jump if equal/Jump if zero			Flags	O D I T S Z A P C
Operands		Clocks	Transfers*	Bytes	Coding Example	
short-label		16 or 4	—	2	JZ ZERO	

JG/JNLE		JG/JNLE short-label Jump if greater/Jump if not less nor equal			Flags	O D I T S Z A P C
Operands		Clocks	Transfers*	Bytes	Coding Example	
short-label		16 or 4	—	2	JG GREATER	

JGE/JNL		JGE/JNL short-label Jump if greater or equal/Jump if not less			Flags	O D I T S Z A P C
Operands		Clocks	Transfers*	Bytes	Coding Example	
short-label		16 or 4	—	2	JGE GREATER__EQUAL	

JL/JNGE		JL/JNGE short-label Jump if less/Jump if not greater nor equal			Flags	O D I T S Z A P C
Operands		Clocks	Transfers*	Bytes	Coding Example	
short-label		16 or 4	—	2	JL LESS	

*For the 8086, add four clocks for each 16-bit word transfer with an odd address. For the 8088, add four clocks for each 16-bit word transfer.

Table 2-21. Instruction Set Reference Data (Cont'd.)

JLE/JNG	JLE/JNG short-label Jump if less or equal/Jump if not greater			Flags	O D I T S Z A P C
Operands	Clocks	Transfers*	Bytes	Coding Example	
short-label	16 or 4	—	2	JNG NOT GREATER	

JMP	JMP target Jump			Flags	O D I T S Z A P C
Operands	Clocks	Transfers*	Bytes	Coding Example	
short-label	15	—	2	JMP SHORT	
near-label	15	—	3	JMP WITHIN SEGMENT	
far-label	15	—	5	JMP FAR LABEL	
memptr16	18 + EA	1	2-4	JMP [BX].TARGET	
regptr16	11	—	2	JMP CX	
memptr32	24 + EA	2	2-4	JMP OTHER.SEG [SI]	

JNC	JNC short-label Jump if not carry			Flags	O D I T S Z A P C
Operands	Clocks	Transfers*	Bytes	Coding Example	
short-label	16 or 4	—	2	JNC NOT_CARRY	

JNE/JNZ	JNE/JNZ short-label Jump if not equal/Jump if not zero			Flags	O D I T S Z A P C
Operands	Clocks	Transfers*	Bytes	Coding Example	
short-label	16 or 4	—	2	JNE NOT_EQUAL	

JNO	JNO short-label Jump if not overflow			Flags	O D I T S Z A P C
Operands	Clocks	Transfers*	Bytes	Coding Example	
short-label	16 or 4	—	2	JNO NO_OVERFLOW	

JNP/JPO	JNP/JPO short-label Jump if not parity/Jump if parity odd			Flags	O D I T S Z A P C
Operands	Clocks	Transfers*	Bytes	Coding Example	
short-label	16 or 4	—	2	JPO ODD_PARITY	

JNS	JNS short-label Jump if not sign			Flags	O D I T S Z A P C
Operands	Clocks	Transfers*	Bytes	Coding Example	
short-label	16 or 4	—	2	JNS POSITIVE	

*For the 8086, add four clocks for each 16-bit word transfer with an odd address. For the 8088, add four clocks for each 16-bit word transfer.

Table 2-21. Instruction Set Reference Data (Cont'd.)

JO		JO short-label Jump if overflow			Flags	O D I T S Z A P C
Operands		Clocks	Transfers*	Bytes	Coding Example	
short-label		16 or 4	—	2	JO SIGNED__OVRFLW	

JP/JPE		JP/JPE short-label Jump if parity/Jump if parity even			Flags	O D I T S Z A P C
Operands		Clocks	Transfers*	Bytes	Coding Example	
short-label		16 or 4	—	2	JPE EVEN__PARITY	

JS		JS short-label Jump if sign			Flags	O D I T S Z A P C
Operands		Clocks	Transfers*	Bytes	Coding Example	
short-label		16 or 4	—	2	JS NEGATIVE	

LAHF		LAHF (no operands) Load AH from flags			Flags	O D I T S Z A P C
Operands		Clocks	Transfers*	Bytes	Coding Example	
(no operands)		4	—	1	LAHF	

LDS		LDS destination,source Load pointer using DS			Flags	O D I T S Z A P C
Operands		Clocks	Transfers	Bytes	Coding Example	
reg16, mem32		16 + EA	2	2-4	LDS SI,DATA.SEG [DI]	

LEA		LEA destination,source Load effective address			Flags	O D I T S Z A P C
Operands		Clocks	Transfers*	Bytes	Coding Example	
reg16, mem16		2 + EA	—	2-4	LEA BX, [BP] [DI]	

LES		LES destination,source Load pointer using ES			Flags	O D I T S Z A P C
Operands		Clocks	Transfers*	Bytes	Coding Example	
reg16, mem32		16 + EA	2	2-4	LES DI, [BX].TEXT BUFF	

*For the 8086, add four clocks for each 16-bit word transfer with an odd address. For the 8088, add four clocks for each 16-bit word transfer.

Table 2-21. Instruction Set Reference Data (Cont'd.)

LOCK	LOCK (no operands) Lock bus			Flags O D I T S Z A P C	
Operands	Clocks	Transfers*	Bytes	Coding Example	
(no operands)	2	—	1	LOCK XCHG FLAG,AL	

LODS	LODS source-string Load string			Flags O D I T S Z A P C	
Operands	Clocks	Transfers*	Bytes	Coding Example	
source-string (repeat) source-string	12 9 + 13/rep	1 1/rep	1 1	LODS CUSTOMER__NAME REP LODS NAME	

LOOP	LOOP short-label Loop			Flags O D I T S Z A P C	
Operands	Clocks	Transfers*	Bytes	Coding Example	
short-label	17/5	—	2	LOOP AGAIN	

LOOPE/LOOPZ	LOOPE/LOOPZ short-label Loop if equal/Loop if zero			Flags O D I T S Z A P C	
Operands	Clocks	Transfers*	Bytes	Coding Example	
short-label	18 or 6	—	2	LOOPE AGAIN	

LOOPNE/LOOPNZ	LOOPNE/LOOPNZ short-label Loop if not equal/Loop if not zero			Flags O D I T S Z A P C	
Operands	Clocks	Transfers*	Bytes	Coding Example	
short-label	19 or 5	—	2	LOOPNE AGAIN	

NMI†	NMI (external nonmaskable interrupt) Interrupt if NMI = 1			Flags O S I T S Z A P C 0 0	
Operands	Clocks	Transfers*	Bytes	Coding Example	
(no operands)	50	5	N/A	N/A	

*For the 8086, add four clocks for each 16-bit word transfer with an odd address. For the 8088, add four clocks for each 16-bit word transfer.

†NMI is not an instruction; it is included in table 2-21 only for timing information.

Table 2-21. Instruction Set Reference Data (Cont'd.)

MOV	MOV destination,source Move			Flags	O D I T S Z A P C
Operands	Clocks	Transfers*	Bytes	Coding Example	
memory, accumulator	10	1	3	MOV ARRAY [SI], AL	
accumulator, memory	10	1	3	MOV AX, TEMP__RESULT	
register, register	2	—	2	MOV AX,CX	
register, memory	8 + EA	1	2-4	MOV BP, STACK__TOP	
memory, register	9 + EA	1	2-4	MOV COUNT [DI], CX	
register, immediate	4	—	2-3	MOV CL, 2	
memory, immediate	10 + EA	1	3-6	MOV MASK [BX] [SI], 2CH	
seg-reg, reg16	2	—	2	MOV ES, CX	
seg-reg, mem16	8 + EA	1	2-4	MOV DS, SEGMENT__BASE	
reg16, seg-reg	2	—	2	MOV BP, SS	
memory, seg-reg	9 + EA	1	2-4	MOV [BX].SEG SAVE, CS	

MOVS	MOVS dest-string,source-string Move string			Flags	O D I T S Z A P C
Operands	Clocks	Transfers*	Bytes	Coding Example	
dest-string, source-string	18	2	1	MOVS LINE EDIT__DATA	
(repeat) dest-string, source-string	9 + 17/rep	2/rep	1	REP MOVS SCREEN, BUFFER	

MOVSB/MOVSW	MOVSB/MOVSW (no operands) Move string (byte/word)			Flags	O D I T S Z A P C
Operands	Clocks	Transfers*	Bytes	Coding Example	
(no operands)	18	2	1	MOVSB	
(repeat) (no operands)	9 + 17/rep	2/rep	1	REP MOVSW	

MUL	MUL source Multiplication, unsigned			Flags	O D I T S Z A P C X U U U U X
Operands	Clocks	Transfers*	Bytes	Coding Example	
reg8	70-77	—	2	MUL BL	
reg16	118-133	—	2	MUL CX	
mem8	(76-83) + EA	1	2-4	MUL MONTH [SI]	
mem16	(124-139) + EA	1	2-4	MUL BAUD__RATE	

*For the 8086, add four clocks for each 16-bit word transfer with an odd address. For the 8088, add four clocks for each 16-bit word transfer.

Table 2-21. Instruction Set Reference Data (Cont'd.)

NEG	NEG destination Negate				Flags	O D I T S Z A P C X X X X X 1*
Operands	**Clocks**	**Transfers***	**Bytes**		**Coding Example**	
register	3	—	2		NEG AL	
memory	16 + EA	2	2-4		NEG MULTIPLIER	

*0 if destination = 0

NOP	NOP (no operands) No Operation				Flags	O D I T S Z A P C
Operands	**Clocks**	**Transfers***	**Bytes**		**Coding Example**	
(no operands)	3	—	1		NOP	

NOT	NOT destination Logical not				Flags	O D I T S Z A P C
Operands	**Clocks**	**Transfers***	**Bytes**		**Coding Example**	
register	3	—	2		NOT AX	
memory	16 + EA	2	2-4		NOT CHARACTER	

OR	OR destination,source Logical inclusive or				Flags	O D I T S Z A P C 0 X X U X 0
Operands	**Clocks**	**Transfers***	**Bytes**		**Coding Example**	
register, register	3	—	2		OR AL, BL	
register, memory	9 + EA	1	2-4		OR DX, PORT_ID [DI]	
memory, register	16 + EA	2	2-4		OR FLAG_BYTE, CL	
accumulator, immediate	4	—	2-3		OR AL, 01101100B	
register, immediate	4	—	3-4		OR CX,01H	
memory, immediate	17 + EA	2	3-6		OR [BX].CMD WORD,0CFH	

OUT	OUT port,accumulator Output byte or word				Flags	O D I T S Z A P C
Operands	**Clocks**	**Transfers***	**Bytes**		**Coding Example**	
immed8, accumulator	10	1	2		OUT 44, AX	
DX, accumulator	8	1	1		OUT DX, AL	

POP	POP destination Pop word off stack				Flags	O D I T S Z A P C
Operands	**Clocks**	**Transfers***	**Bytes**		**Coding Example**	
register	8	1	1		POP DX	
seg-reg (CS illegal)	8	1	1		POP DS	
memory	17 + EA	2	2-4		POP PARAMETER	

*For the 8086, add four clocks for each 16-bit word transfer with an odd address. For the 8088, add four clocks for each 16-bit word transfer

Table 2-21. Instruction Set Reference Data (Cont'd.)

POPF	POPF (no operands) Pop flags off stack			Flags	O D I T S Z A P C R R R R R R R R R
Operands	**Clocks**	**Transfers***	**Bytes**	**Coding Example**	
(no operands)	8	1	1	POPF	

PUSH	PUSH source Push word onto stack			Flags	O D I T S Z A P C
Operands	**Clocks**	**Transfers***	**Bytes**	**Coding Example**	
register	11	1	1	PUSH SI	
seg-reg (CS legal)	10	1	1	PUSH ES	
memory	16 + EA	2	2-4	PUSH RETURN__CODE [SI]	

PUSHF	PUSHF (no operands) Push flags onto stack			Flags	O D I T S Z A P C
Operands	**Clocks**	**Transfers***	**Bytes**	**Coding Example**	
(no operands)	10	1	1	PUSHF	

RCL	RCL destination,count Rotate left through carry			Flags	O D I T S Z A P C X X
Operands	**Clocks**	**Transfers***	**Bytes**	**Coding Example**	
register, 1	2	—	2	RCL CX, 1	
register, CL	8 + 4/bit	—	2	RCL AL, CL	
memory, 1	15 + EA	2	2-4	RCL ALPHA, 1	
memory, CL	20 + EA + 4/bit	2	2-4	RCL [BP].PARM, CL	

RCR	RCR designation,count Rotate right through carry			Flags	O D I T S Z A P C X X
Operands	**Clocks**	**Transfers***	**Bytes**	**Coding Example**	
register, 1	2	—	2	RCR BX, 1	
register, CL	8 + 4/bit	—	2	RCR BL, CL	
memory, 1	15 + EA	2	2-4	RCR [BX] STATUS, 1	
memory, CL	20 + EA + 4/bit	2	2-4	RCR ARRAY [DI], CL	

REP	REP (no operands) Repeat string operation			Flags	O D I T S Z A P C
Operands	**Clocks**	**Transfers***	**Bytes**	**Coding Example**	
(no operands)	2	—	1	REP MOVS DEST, SRCE	

*For the 8086, add four clocks for each 16-bit word transfer with an odd address. For the 8088, add four clocks for each 16-bit word transfer.

Table 2-21. Instruction Set Reference Data (Cont'd.)

REPE/REPZ

REPE/REPZ (no operands) Repeat string operation while equal/while zero				Flags	O D I T S Z A P C
Operands	Clocks	Transfers*	Bytes	Coding Example	
(no operands)	2	—	1	REPE CMPS DATA, KEY	

REPNE/REPNZ

REPNE/REPNZ (no operands) Repeat string operation while not equal/not zero				Flags	O D I T S Z A P C
Operands	Clocks	Transfers*	Bytes	Coding Example	
(no operands)	2	—	1	REPNE SCAS INPUT LINE	

RET

RET optional-pop-value Return from procedure				Flags	O D I T S Z A P C
Operands	Clocks	Transfers*	Bytes	Coding Example	
(intra-segment. no pop)	8	1	1	RET	
(intra-segment, pop)	12	1	3	RET 4	
(inter-segment, no pop)	18	2	1	RET	
(inter-segment, pop)	17	2	3	RET 2	

ROL

ROL destination,count Rotate left				Flags	O D I T S Z A P C X X
Operands	Clocks	Transfers	Bytes	Coding Examples	
register, 1	2	—	2	ROL BX, 1	
register, CL	8 + 4/bit	—	2	ROL DI, CL	
memory, 1	15 + EA	2	2-4	ROL FLAG _BYTE [DI],1	
memory, CL	20 + EA + 4/bit	2	2-4	ROL ALPHA , CL	

ROR

ROR destination,count Rotate right				Flags	O D I T S Z A P C X X
Operand	Clocks	Transfers*	Bytes	Coding Example	
register, 1	2	—	2	ROR AL, 1	
register, CL	8 + 4/bit	—	2	ROR BX, CL	
memory, 1	15 + EA	2	2-4	ROR PORT_STATUS, 1	
memory, CL	20 + EA + 4/bit	2	2-4	ROR CMD_WORD, CL	

SAHF

SAHF (no operands) Store AH into flags				Flags	O D I T S Z A P C R R R R R
Operands	Clocks	Transfers*	Bytes	Coding Example	
(no operands)	4	—	1	SAHF	

*For the 8086. add four clocks for each 16-bit word transfer with an odd address. For the 8088. add four clocks for each 16-bit word transfer.

Mnemonics   Intel, 1978

Table 2-21. Instruction Set Reference Data (Cont'd.)

SAL/SHL	SAL/SHL destination,count Shift arithmetic left/Shift logical left			Flags	O D I T S Z A P C X X
Operands	**Clocks**	**Transfers***	**Bytes**	**Coding Examples**	
register,1	2	—	2	SAL AL,1	
register, CL	8 + 4/bit	—	2	SHL DI, CL	
memory,1	15 + EA	2	2-4	SHL [BX].OVERDRAW, 1	
memory, CL	20 + EA + 4/bit	2	2-4	SAL STORE__COUNT, CL	

SAR	SAR destination,source Shift arithmetic right			Flags	O D I T S Z A P C X X X U X X
Operands	**Clocks**	**Transfers***	**Bytes**	**Coding Example**	
register, 1	2	—	2	SAR DX, 1	
register, CL	8 + 4/bit	—	2	SAR DI, CL	
memory, 1	15 + EA	2	2-4	SAR N__BLOCKS, 1	
memory, CL	20 + EA + 4/bit	2	2-4	SAR N__BLOCKS, CL	

SBB	SBB destination,source Subtract with borrow			Flags	O D I T S Z A P C X X X X X X
Operands	**Clocks**	**Transfers***	**Bytes**	**Coding Example**	
register, register	3	—	2	SBB BX, CX	
register, memory	9 + EA	1	2-4	SBB DI, [BX].PAYMENT	
memory, register	16 + EA	2	2-4	SBB BALANCE, AX	
accumulator, immediate	4	—	2-3	SBB AX, 2	
register, immediate	4	—	3-4	SBB CL, 1	
memory, immediate	17 + EA	2	3-6	SBB COUNT [SI], 10	

SCAS	SCAS dest-string Scan string			Flags	O D I T S Z A P C X X X X X X
Operands	**Clocks**	**Transfers***	**Bytes**	**Coding Example**	
dest-string	15	1	1	SCAS INPUT__LINE	
(repeat) dest-string	9 + 15/rep	1/rep	1	REPNE SCAS BUFFER	

SEGMENT†	SEGMENT override prefix Override to specified segment			Flags	O D I T S Z A P C
Operands	**Clocks**	**Transfers***	**Bytes**	**Coding Example**	
(no operands)	2	—	1	MOV SS:PARAMETER, AX	

*For the 8086, add four clocks for each 16-bit word transfer with an odd address. For the 8088, add four clocks for each 16-bit word transfer.

†ASM-86 incorporates the segment override prefix into the operand specification and not as a separate instruction. SEGMENT is included in table 2-21 only for timing information.

Appendix H

Table 2-21. Instruction Set Reference Data (Cont'd.)

SHR

SHR destination,count				Flags	O D I T S Z A P C
Shift logical right					X X

Operands	Clocks	Transfers*	Bytes	Coding Example
register, 1	2	—	2	SHR SI, 1
register, CL	8 + 4/bit	—	2	SHR SI, CL
memory, 1	15 + EA	2	2-4	SHR ID....BYTE [SI] [BX], 1
memory, CL	20 + EA + 4/bit	2	2-4	SHR INPUT WORD, CL

SINGLE STEP†

SINGLE STEP (Trap flag interrupt)				Flags	O D I T S Z A P C
Interrupt if TF = 1					0 0

Operands	Clocks	Transfers*	Bytes	Coding Example
(no operands)	50	5	N/A	N/A

STC

STC (no operands)				Flags	O D I T S Z A P C
Set carry flag					1

Operands	Clocks	Transfers*	Bytes	Coding Example
(no operands)	2	—	1	STC

STD

STD (no operands)				Flags	O D I T S Z A P C
Set direction flag					1

Operands	Clocks	Transfers*	Bytes	Coding Example
(no operands)	2	—	1	STD

STI

STI (no operands)				Flags	O D I T S Z A P C
Set interrupt enable flag					1

Operands	Clocks	Transfers*	Bytes	Coding Example
(no operands)	2	—	1	STI

STOS

STOS dest-string				Flags	O D I T S Z A P C
Store byte or word string					

Operands	Clocks	Transfers*	Bytes	Coding Example
dest-string	11	1	1	STOS PRINT__LINE
(repeat) dest-string	9 + 10/rep	1/rep	1	REP STOS DISPLAY

*For the 8086, add four clocks for each 16-bit word transfer with an odd address. For the 8088, add four clocks for each 16-bit word transfer.

†SINGLE STEP is not an instruction; it is included in table 2-21 only for timing information

Table 2-21. Instruction Set Reference Data (Cont'd.)

SUB					SUB destination,source Subtraction		Flags	O D I T S Z A P C X X X X X X
Operands		**Clocks**	**Transfers***	**Bytes**			**Coding Example**	
register, register		3	—	2			SUB CX, BX	
register, memory		9 + EA	1	2-4			SUB DX, MATH__TOTAL [SI]	
memory, register		16 + EA	2	2-4			SUB [BP + 2], CL	
accumulator, immediate		4	—	2-3			SUB AL, 10	
register, immediate		4	—	3-4			SUB SI, 5280	
memory, immediate		17 + EA	2	3-6			SUB [BP].BALANCE, 1000	

TEST					TEST destination,source Test or non-destructive logical and		Flags	O D I T S Z A P C 0 X X U X 0
Operands		**Clocks**	**Transfers***	**Bytes**			**Coding Example**	
register, register		3	—	2			TEST SI, DI	
register, memory		9 + EA	1	2-4			TEST SI, END__COUNT	
accumulator, immediate		4	—	2-3			TEST AL, 00100000B	
register, immediate		5	—	3-4			TEST BX, 0CC4H	
memory, immediate		11 + EA	—	3-6			TEST RETURN__CODE, 01H	

WAIT					WAIT (no operands) Wait while $\overline{TEST}$ pin not asserted		Flags	O D I T S Z A P C
Operands		**Clocks**	**Transfers***	**Bytes**			**Coding Example**	
(no operands)		3 + 5n	—	1			WAIT	

XCHG					XCHG destination,source Exchange		Flags	O D I T S Z A P C
Operands		**Clocks**	**Transfers***	**Bytes**			**Coding Example**	
accumulator, reg16		3	—	1			XCHG AX, BX	
memory, register		17 + EA	2	2-4			XCHG SEMAPHORE, AX	
register, register		4	—	2			XCHG AL, BL	

XLAT					XLAT source-table Translate		Flags	O D I T S Z A P C
Operands		**Clocks**	**Transfers***	**Bytes**			**Coding Example**	
source-table		11	1	1			XLAT ASCII TAB	

*For the 8086, add four clocks for each 16-bit word transfer with an odd address. For the 8088, add four clocks for each 16-bit word transfer.

Table 2-21. Instruction Set Reference Data (Cont'd.)

XOR	XOR destination, source Logical exclusive or			Flags	O D I T S Z A P C 0 X X U X 0
Operands	**Clocks**	**Transfers***	**Bytes**	**Coding Example**	
register, register	3	—	2	XOR CX, BX	
register, memory	9 + EA	1	2-4	XOR CL, MASK BYTE	
memory, register	16 + EA	2	2-4	XOR ALPHA [SI], DX	
accumulator, immediate	4	—	2-3	XOR AL, 01000010B	
register, immediate	4	—	3-4	XOR SI, 00C2H	
memory, immediate	17 + EA	2	3-6	XOR RETURN CODE, 0D2H	

*For the 8086, add four clocks for each 16-bit word transfer with an odd address. For the 8088, add four clocks for each 16-bit word transfer.

Table 4-12. 8086 Instruction Encoding

DATA TRANSFER

MOV = Move:

	7 6 5 4 3 2 1 0	7 6 5 4 3 2 1 0	7 6 5 4 3 2 1 0	7 6 5 4 3 2 1 0	7 6 5 4 3 2 1 0	7 6 5 4 3 2 1 0
Register/memory to/from register	1 0 0 0 1 0 d w	mod reg r/m	(DISP-LO)	(DISP-HI)		
Immediate to register/memory	1 1 0 0 0 1 1 w	mod 0 0 0 r/m	(DISP-LO)	(DISP-HI)	data	data if w = 1
Immediate to register	1 0 1 1 w reg	data	data if w = 1			
Memory to accumulator	1 0 1 0 0 0 0 w	addr-lo	addr-hi			
Accumulator to memory	1 0 1 0 0 0 1 w	addr-lo	addr-hi			
Register/memory to segment register	1 0 0 0 1 1 1 0	mod 0 SR r/m	(DISP-LO)	(DISP-HI)		
Segment register to register/memory	1 0 0 0 1 1 0 0	mod 0 SR r/m	(DISP-LO)	(DISP-HI)		

PUSH = Push:

Register/memory	1 1 1 1 1 1 1 1	mod 1 1 0 r/m	(DISP-LO)	(DISP-HI)
Register	0 1 0 1 0 reg			
Segment register	0 0 0 reg 1 1 0			

POP = Pop:

Register/memory	1 0 0 0 1 1 1 1	mod 0 0 0 r/m	(DISP-LO)	(DISP-HI)
Register	0 1 0 1 1 reg			
Segment register	0 0 0 reg 1 1 1			

Mnemonics ⓒ Intel, 1978

Table 4-12. 8086 Instruction Encoding (Cont'd.)

DATA TRANSFER (Cont'd.)

XCHG = Exchange:

	7 6 5 4 3 2 1 0	7 6 5 4 3 2 1 0	7 6 5 4 3 2 1 0	7 6 5 4 3 2 1 0	7 6 5 4 3 2 1 0	7 6 5 4 3 2 1 0
Register/memory with register	1 0 0 0 0 1 1 w	mod reg r/m	(DISP-LO)	(DISP-HI)		
Register with accumulator	1 0 0 1 0 reg					

IN = Input from:

Fixed port	1 1 1 0 0 1 0 w	DATA-8	
Variable port	1 1 1 0 1 1 0 w		

OUT = Output to:

Fixed port	1 1 1 0 0 1 1 w	DATA-8			
Variable port	1 1 1 0 1 1 1 w				
XLAT = Translate byte to AL	1 1 0 1 0 1 1 1				
LEA = Load EA to register	1 0 0 0 1 1 0 1	mod reg r/m	(DISP-LO)	(DISP-HI)	
LDS = Load pointer to DS	1 1 0 0 0 1 0 1	mod reg r/m	(DISP-LO)	(DISP-HI)	
LES = Load pointer to ES	1 1 0 0 0 1 0 0	mod reg r/m	(DISP-LO)	(DISP-HI)	
LAHF = Load AH with flags	1 0 0 1 1 1 1 1				
SAHF = Store AH into flags	1 0 0 1 1 1 1 0				
PUSHF = Push flags	1 0 0 1 1 1 0 0				
POPF = Pop flags	1 0 0 1 1 1 0 1				

ARITHMETIC

ADD = Add:

Reg/memory with register to either	0 0 0 0 0 0 d w	mod reg r/m	(DISP-LO)	(DISP-HI)		
Immediate to register/memory	1 0 0 0 0 0 s w	mod 0 0 0 r/m	(DISP-LO)	(DISP-HI)	data	data if s: w=01
Immediate to accumulator	0 0 0 0 0 1 0 w	data	data if w=1			

ADC = Add with carry:

Reg/memory with register to either	0 0 0 1 0 0 d w	mod reg r/m	(DISP-LO)	(DISP-HI)		
Immediate to register/memory	1 0 0 0 0 0 s w	mod 0 1 0 r/m	(DISP-LO)	(DISP-HI)	data	data if s: w=01
Immediate to accumulator	0 0 0 1 0 1 0 w	data	data if w=1			

INC = Increment:

Register/memory	1 1 1 1 1 1 1 w	mod 0 0 0 r/m	(DISP-LO)	(DISP-HI)
Register	0 1 0 0 0 reg			
AAA = ASCII adjust for add	0 0 1 1 0 1 1 1			
DAA = Decimal adjust for add	0 0 1 0 0 1 1 1			

Table 4-12. 8086 Instruction Encoding (Cont'd.)

ARITHMETIC (Cont'd.)

	76543210	76543210	76543210	76543210	76543210	76543210
SUB = Subtract:						
Reg/memory and register to either	001010 d w	mod reg r/m	(DISP-LO)	(DISP-HI)		
Immediate from register/memory	100000 s w	mod 1 0 1 r/m	(DISP-LO)	(DISP-HI)	data	data if s: w=01
Immediate from accumulator	0010110 w	data	data if w=1			
SBB = Subtract with borrow:						
Reg/memory and register to either	000110 d w	mod reg r/m	(DISP-LO)	(DISP-HI)		
Immediate from register/memory	100000 s w	mod 0 1 1 r/m	(DISP-LO)	(DISP-HI)	data	data if s: w=01
Immediate from accumulator	0001110 w	data	data if w=1			
DEC Decrement:						
Register/memory	1111111 w	mod 0 0 1 r/m	(DISP-LO)	(DISP-HI)		
Register	0 1 0 0 1 reg					
NEG Change sign	1111011 w	mod 0 1 1 r/m	(DISP-LO)	(DISP-HI)		
CMP = Compare:						
Register/memory and register	001110 d w	mod reg r/m	(DISP-LO)	(DISP-HI)		
Immediate with register/memory	100000 s w	mod 1 1 1 r/m	(DISP-LO)	(DISP-HI)	data	data if s: w=1
Immediate with accumulator	0011110 w	data				
AAS ASCII adjust for subtract	00111111					
DAS Decimal adjust for subtract	00101111					
MUL Multiply (unsigned)	1111011 w	mod 1 0 0 r/m	(DISP-LO)	(DISP-HI)		
IMUL Integer multiply (signed)	1111011 w	mod 1 0 1 r/m	(DISP-LO)	(DISP-HI)		
AAM ASCII adjust for multiply	11010100	00001010	(DISP-LO)	(DISP-HI)		
DIV Divide (unsigned)	1111011 w	mod 1 1 0 r/m	(DISP-LO)	(DISP-HI)		
IDIV Integer divide (signed)	1111011 w	mod 1 1 1 r/m	(DISP-LO)	(DISP-HI)		
AAD ASCII adjust for divide	11010101	00001010	(DISP-LO)	(DISP-HI)		
CBW Convert byte to word	10011000					
CWD Convert word to double word	10011001					
LOGIC						
NOT Invert	1111011 w	mod 0 1 0 r/m	(DISP-LO)	(DISP-HI)		
SHL/SAL Shift logical/arithmetic left	110100 v w	mod 1 0 0 r/m	(DISP-LO)	(DISP-HI)		
SHR Shift logical right	110100 v w	mod 1 0 1 r/m	(DISP-LO)	(DISP-HI)		
SAR Shift arithmetic right	110100 v w	mod 1 1 1 r/m	(DISP-LO)	(DISP-HI)		
ROL Rotate left	110100 v w	mod 0 0 0 r/m	(DISP-LO)	(DISP-HI)		

Mnemonics © Intel, 1978

Table 4-12. 8086 Instruction Encoding (Cont'd.)

LOGIC (Cont'd.)

	7 6 5 4 3 2 1 0	7 6 5 4 3 2 1 0	7 6 5 4 3 2 1 0	7 6 5 4 3 2 1 0	7 6 5 4 3 2 1 0	7 6 5 4 3 2 1 0
ROR Rotate right	1 1 0 1 0 0 v w	mod 0 0 1 r/m	(DISP-LO)	(DISP-HI)		
RCL Rotate through carry flag left	1 1 0 1 0 0 v w	mod 0 1 0 r/m	(DISP-LO)	(DISP-HI)		
RCR Rotate through carry right	1 1 0 1 0 0 v w	mod 0 1 1 r/m	(DISP-LO)	(DISP-HI)		

AND = And:

Reg/memory with register to either	0 0 1 0 0 0 d w	mod reg r/m	(DISP-LO)	(DISP-HI)		
Immediate to register/memory	1 0 0 0 0 0 0 w	mod 1 0 0 r/m	(DISP-LO)	(DISP-HI)	data	data if w=1
Immediate to accumulator	0 0 1 0 0 1 0 w	data	data if w=1			

TEST = And function to flags no result:

Register/memory and register	0 0 0 1 0 0 d w	mod reg r/m	(DISP-LO)	(DISP-HI)		
Immediate data and register/memory	1 1 1 1 0 1 1 w	mod 0 0 0 r/m	(DISP-LO)	(DISP-HI)	data	data if w=1
Immediate data and accumulator	1 0 1 0 1 0 0 w	data				

OR = Or:

Reg/memory and register to either	0 0 0 0 1 0 d w	mod reg r/m	(DISP-LO)	(DISP-HI)		
Immediate to register/memory	1 0 0 0 0 0 0 w	mod 0 0 1 r/m	(DISP-LO)	(DISP-HI)	data	data if w=1
Immediate to accumulator	0 0 0 0 1 1 0 w	data	data if w=1			

XOR = Exclusive or:

Reg/memory and register to either	0 0 1 1 0 0 d w	mod reg r/m	(DISP-LO)	(DISP-HI)		
Immediate to register/memory	0 0 1 1 0 1 0 w	data	(DISP-LO)	(DISP-HI)	data	data if w=1
Immediate to accumulator	0 0 1 1 0 1 0 w	data	data if w=1			

STRING MANIPULATION

REP = Repeat	1 1 1 1 0 0 1 z
MOVS = Move byte/word	1 0 1 0 0 1 0 w
CMPS = Compare byte/word	1 0 1 0 0 1 1 w
SCAS = Scan byte/word	1 0 1 0 1 1 1 w
LODS = Load byte/wd to AL/AX	1 0 1 0 1 1 0 w
STDS = Stor byte/wd from AL/A	1 0 1 0 1 0 1 w

Table 4-12. 8086 Instruction Encoding (Cont'd.)

CONTROL TRANSFER

CALL = Call:

	7 6 5 4 3 2 1 0	7 6 5 4 3 2 1 0	7 6 5 4 3 2 1 0	7 6 5 4 3 2 1 0	7 6 5 4 3 2 1 0	7 6 5 4 3 2 1 0
Direct within segment	1 1 1 0 1 0 0 0	IP-INC-LO	IP-INC-HI			
Indirect within segment	1 1 1 1 1 1 1 1	mod 0 1 0 r/m	(DISP-LO)	(DISP-HI)		
Direct intersegment	1 0 0 1 1 0 1 0	IP-lo	IP-hi			
		CS-lo	CS-hi			
Indirect intersegment	1 1 1 1 1 1 1 1	mod 0 1 1 r/m	(DISP-LO)	(DISP-HI)		

JMP = Unconditional Jump:

Direct within segment	1 1 1 0 1 0 0 1	IP-INC-LO	IP-INC-HI	
Direct within segment-short	1 1 1 0 1 0 1 1	IP-INC8		
Indirect within segment	1 1 1 1 1 1 1 1	mod 1 0 0 r/m	(DISP-LO)	(DISP-HI)
Direct intersegment	1 1 1 0 1 0 1 0	IP-lo	IP-hi	
		CS-lo	CS-hi	
Indirect intersegment	1 1 1 1 1 1 1 1	mod 1 0 1 r/m	(DISP-LO)	(DISP-HI)

RET = Return from CALL:

Within segment	1 1 0 0 0 0 1 1		
Within seg adding immed to SP	1 1 0 0 0 0 1 0	data-lo	data-hi
Intersegment	1 1 0 0 1 0 1 1		
Intersegment adding immediate to SP	1 1 0 0 1 0 1 0	data-lo	data-hi
JE/JZ = Jump on equal/zero	0 1 1 1 0 1 0 0	IP-INC8	
JL/JNGE = Jump on less/not greater or equal	0 1 1 1 1 1 0 0	IP-INC8	
JLE/JNG = Jump on less or equal/not greater	0 1 1 1 1 1 1 0	IP-INC8	
JB/JNAE = Jump on below/not above or equal	0 1 1 1 0 0 1 0	IP-INC8	
JBE/JNA = Jump on below or equal/not above	0 1 1 1 0 1 1 0	IP-INC8	
JP/JPE = Jump on parity/parity even	0 1 1 1 1 0 1 0	IP-INC8	
JO = Jump on overflow	0 1 1 1 0 0 0 0	IP-INC8	
JS = Jump on sign	0 1 1 1 1 0 0 0	IP-INC8	
JNE/JNZ = Jump on not equal/not zer0	0 1 1 1 0 1 0 1	IP-INC8	
JNL/JGE = Jump on not less/greater or equal	0 1 1 1 1 1 0 1	IP-INC8	
JNLE/JG = Jump on not less or equal/greater	0 1 1 1 1 1 1 1	IP-INC8	
JNB/JAE = Jump on not below/above or equal	0 1 1 1 0 0 1 1	IP-INC8	
JNBE/JA = Jump on not below or equal/above	0 1 1 1 0 1 1 1	IP-INC8	
JNP/JPO = Jump on not par/par odd	0 1 1 1 1 0 1 1	IP-INC8	
JNO = Jump on not overflow	0 1 1 1 0 0 0 1	IP-INC8	

Table 4-12. 8086 Instruction Encoding (Cont'd.)

CONTROL TRANSFER (Cont'd.)

RET = Return from CALL:

	7 6 5 4 3 2 1 0	7 6 5 4 3 2 1 0	7 6 5 4 3 2 1 0	7 6 5 4 3 2 1 0	7 6 5 4 3 2 1 0	7 6 5 4 3 2 1 0

JNS = Jump on not sign	0 1 1 1 1 0 0 1	IP-INC8
LOOP = Loop CX times	1 1 1 0 0 0 1 0	IP-INC8
LOOPZ/LOOPE = Loop while zero/equal	1 1 1 0 0 0 0 1	IP-INC8
LOOPNZ/LOOPNE = Loop while not zero/equal	1 1 1 0 0 0 0 0	IP-INC8
JCXZ = Jump on CX zero	1 1 1 0 0 0 1 1	IP-INC8

INT = Interrupt:

Type specified	1 1 0 0 1 1 0 1	DATA-8
Type 3	1 1 0 0 1 1 0 0	
INTO = Interrupt on overflow	1 1 0 0 1 1 1 0	
IRET = Interrupt return	1 1 0 0 1 1 1 1	

PROCESSOR CONTROL

CLC = Clear carry	1 1 1 1 1 0 0 0			
CMC = Complement carry	1 1 1 1 0 1 0 1			
STC = Set carry	1 1 1 1 1 0 0 1			
CLD = Clear direction	1 1 1 1 1 1 0 0			
STD = Set direction	1 1 1 1 1 1 0 1			
CLI = Clear interrupt	1 1 1 1 1 0 1 0			
STI = Set interrupt	1 1 1 1 1 0 1 1			
HLT = Halt	1 1 1 1 0 1 0 0			
WAIT = Wait	1 0 0 1 1 0 1 1			
ESC = Escape (to external device)	1 1 0 1 1 x x x	mod y y y r/m	(DISP-LO)	(DISP-HI)
LOCK = Bus lock prefix	1 1 1 1 0 0 0 0			
SEGMENT = Override prefix	0 0 1 reg 1 1 0			

Table 4-13. Machine Instruction Decoding Guide

1ST BYTE		2ND BYTE	BYTES 3, 4, 5, 6	ASM-86 INSTRUCTION FORMAT	
HEX	BINARY				
00	0000 0000	MOD REG R/M	(DISP-LO),(DISP-HI)	ADD	REG8/MEM8,REG8
01	0000 0001	MOD REG R/M	(DISP-LO),(DISP-HI)	ADD	REG16/MEM16,REG16
02	0000 0010	MOD REG R/M	(DISP-LO),(DISP-HI)	ADD	REG8,REG8/MEM8
03	0000 0011	MOD REG R/M	(DISP-LO),(DISP-HI)	ADD	REG16,REG16/MEM16
04	0000 0100	DATA-8		ADD	AL,IMMED8
05	0000 0101	DATA-LO	DATA-HI	ADD	AX,IMMED16
06	0000 0110			PUSH	ES
07	0000 0111			POP	ES

Table 4-13. Machine Instruction Decoding Guide (Cont'd.)

| 1ST BYTE | | 2ND BYTE | BYTES 3,4,5,6 | ASM-86 INSTRUCTION FORMAT | |
HEX	BINARY				
08	0000 1000	MOD REG R/M	(DISP-LO),(DISP-HI)	OR	REG8/MEM8,REG8
09	0000 1001	MOD REG R/M	(DISP-LO),(DISP-HI)	OR	REG16/MEM16,REG16
0A	0000 1010	MOD REG R/M	(DISP-LO),(DISP-HI)	OR	REG8,REG8/MEM8
0B	0000 1011	MOD REG R/M	(DISP-LO),(DISP-HI)	OR	REG16,REG16/MEM16
0C	0000 1100	DATA-8		OR	AL,IMMED8
0D	0000 1101	DATA-LO	DATA-HI	OR	AX,IMMED16
0E	0000 1110			PUSH	CS
0F	0000 1111			(not used)	
10	0001 0000	MOD REG R/M	(DISP-LO),(DISP-HI)	ADC	REG8/MEM8,REG8
11	0001 0001	MOD REG R/M	(DISP-LO),(DISP-HI)	ADC	REG16/MEM16,REG16
12	0001 0010	MOD REG R/M	(DISP-LO),(DISP-HI)	ADC	REG8,REG8/MEM8
13	0001 0011	MOD REG R/M	(DISP-LO),(DISP-HI)	ADC	REG16,REG16/MEM16
14	0001 0100	DATA-8		ADC	AL,IMMED8
15	0001 0101	DATA-LO	DATA-HI	ADC	AX,IMMED16
16	0001 0110			PUSH	SS
17	0001 0111			POP	SS
18	0001 1000	MOD REG R/M	(DISP-LO),(DISP-HI)	SBB	REG8/MEM8,REG8
19	0001 1001	MOD REG R/M	(DISP-LO),(DISP-HI)	SBB	REG16/MEM16,REG16
1A	0001 1010	MOD REG R/M	(DISP-LO),(DISP-HI)	SBB	REG8,REG8/MEM8
1B	0001 1011	MOD REG R/M	(DISP-LO),(DISP-HI)	SBB	REG16,REG16/MEM16
1C	0001 1100	DATA-8		SBB	AL,IMMED8
1D	0001 1101	DATA-LO	DATA-HI	SBB	AX,IMMED16
1E	0001 1110			PUSH	DS
1F	0001 1111			POP	DS
20	0010 0000	MOD REG R/M	(DISP-LO),(DISP-HI)	AND	REG8/MEM8,REG8
21	0010 0001	MOD REG R/M	(DISP-LO),(DISP-HI)	AND	REG16/MEM16,REG16
22	0010 0010	MOD REG R/M	(DISP-LO),(DISP-HI)	AND	REG8,REG8/MEM8
23	0010 0011	MOD REG R/M	(DISP-LO),(DISP-HI)	AND	REG16,REG16/MEM16
24	0010 0100	DATA-8		AND	AL,IMMED8
25	0010 0101	DATA-LO	DATA-HI	AND	AX,IMMED16
26	0010 0110			ES:	(segment override prefix)
27	0010 0111			DAA	
28	0010 1000	MOD REG R/M	(DISP-LO),(DISP-HI)	SUB	REG8/MEM8,REG8
29	0010 1001	MOD REG R/M	(DISP-LO),(DISP-HI)	SUB	REG16/MEM16,REG16
2A	0010 1010	MOD REG R/M	(DISP-LO),(DISP-HI)	SUB	REG8,REG8/MEM8
2B	0010 1011	MOD REG R/M	(DISP-LO,(DISP-HI)	SUB	REG16,REG16/MEM16
2C	0010 1100	DATA-8		SUB	AL,IMMED8
2D	0010 1101	DATA-LO	DATA-HI	SUB	AX,IMMED16
2E	0010 1110			CS:	(segment override prefix)
2F	0010 1111			DAS	
30	0011 0000	MOD REG R/M	(DISP-LO),(DISP-HI)	XOR	REG8/MEM8,REG8
31	0011 0001	MOD REG R/M	(DISP-LO),(DISP-HI)	XOR	REG16/MEM16,REG16
32	0011 0010	MOD REG R/M	(DISP-LO),(DISP-HI)	XOR	REG8,REG8/MEM8
33	0011 0011	MOD REG R/M	(DISP-LO),(DISP-HI)	XOR	REG16,REG16/MEM16
34	0011 0100	DATA-8		XOR	AL,IMMED8
35	0011 0101	DATA-LO	DATA-HI	XOR	AX,IMMED16
36	0011 0110			SS:	(segment override prefix)

Mnemonics = Intel, 1978

Table 4-13. Machine Instruction Decoding Guide (Cont'd.)

| 1ST BYTE | | 2ND BYTE | BYTES 3,4,5,6 | ASM-86 INSTRUCTION FORMAT | |
HEX	BINARY				
37	0011 0110			AAA	
38	0011 1000	MOD REG R/M	(DISP-LO),(DISP-HI)	CMP	REG8/MEM8,REG8
39	0011 1001	MOD REG R/M	(DISP-LO),(DISP-HI)	CMP	REG16/MEM16,REG16
3A	0011 1010	MOD REG R/M	(DISP-LO),(DISP-HI)	CMP	REG8,REG8/MEM8
3B	0011 1011	MOD REG R/M	(DISP-LO),(DISP-HI)	CMP	REG16,REG16/MEM16
3C	0011 1100	DATA-8		CMP	AL,IMMED8
3D	0011 1101	DATA-LO	DATA-HI	CMP	AX,IMMED16
3E	0011 1110			DS:	(segment override prefix)
3F	0011 1111			AAS	
40	0100 0000			INC	AX
41	0100 0001			INC	CX
42	0100 0010			INC	DX
43	0100 0011			INC	BX
44	0100 0100			INC	SP
45	0100 0101			INC	BP
46	0100 0110			INC	SI
47	0100 0111			INC	DI
48	0100 1000			DEC	AX
49	0100 1001			DEC	CX
4A	0100 1010			DEC	DX
4B	0100 1011			DEC	BX
4C	0100 1100			DEC	SP
4D	0100 1101			DEC	BP
4E	0100 1110			DEC	SI
4F	0100 1111			DEC	DI
50	0101 0000			PUSH	AX
51	0101 0001			PUSH	CX
52	0101 0010			PUSH	DX
53	0101 0011			PUSH	BX
54	0101 0100			PUSH	SP
55	0101 0101			PUSH	BP
56	0101 0110			PUSH	SI
57	0101 0111			PUSH	DI
58	0101 1000			POP	AX
59	0101 1001			POP	CX
5A	0101 1010			POP	DX
5B	0101 1011			POP	BX
5C	0101 1100			POP	SP
5D	0101 1101			POP	BP
5E	0101 1110			POP	SI
5F	0101 1111			POP	DI
60	0110 0000			(not used)	
61	0110 0001			(not used)	
62	0110 0010			(not used)	
63	0110 0011			(not used)	
64	0110 0100			(not used)	
65	0110 0101			(not used)	
66	0110 0110			(not used)	
67	0110 0111			(not used)	

Mnemonics © Intel, 1978

Table 4-13. Machine Instruction Decoding Guide (Cont'd.)

1ST BYTE		2ND BYTE	BYTES 3,4,5,6	ASM-86 INSTRUCTION FORMAT	
HEX	BINARY				
68	0110 1000			(not used)	
69	0110 1001			(not used)	
6A	0110 1010			(not used)	
6B	0110 1011			(not used)	
6C	0110 1100			(not used)	
6D	0110 1101			(not used)	
6E	0110 1110			(not used)	
6F	0110 1111			(not used)	
70	0111 0000	IP-INC8		JO	SHORT-LABEL
71	0111 0001	IP-INC8		JNO	SHORT-LABEL
72	0111 0010	IP-INC8		JB/JNAE/ JC	SHORT-LABEL
73	0111 0011	IP-INC8		JNB/JAE/ JNC	SHORT-LABEL
74	0111 0100	IP-INC8		JE/JZ	SHORT-LABEL
75	0111 0101	IP-INC8		JNE/JNZ	SHORT-LABEL
76	0111 0110	IP-INC8		JBE/JNA	SHORT-LABEL
77	0111 0111	IP-INC8		JNBE/JA	SHORT-LABEL
78	0111 1000	IP-INC8		JS	SHORT-LABEL
79	0111 1001	IP-INC8		JNS	SHORT-LABEL
7A	0111 1010	IP-INC8		JP/JPE	SHORT-LABEL
7B	0111 1011	IP-INC8		JNP/JPO	SHORT-LABEL
7C	0111 1100	IP-INC8		JL/JNGE	SHORT-LABEL
7D	0111 1101	IP-INC8		JNL/JGE	SHORT-LABEL
7E	0111 1110	IP-INC8		JLE/JNG	SHORT-LABEL
7F	0111 1111	IP-INC8		JNLE/JG	SHORT-LABEL
80	1000 0000	MOD 000 R/M	(DISP-LO),(DISP-HI), DATA-8	ADD	REG8/MEM8,IMMED8
80	1000 0000	MOD 001 R/M	(DISP-LO),(DISP-HI), DATA-8	OR	REG8/MEM8,IMMED8
80	1000 0000	MOD 010 R/M	(DISP-LO),(DISP-HI), DATA-8	ADC	REG8/MEM8,IMMED8
80	1000 0000	MOD 011 R/M	(DISP-LO),(DISP-HI), DATA-8	SBB	REG8/MEM8,IMMED8
80	1000 0000	MOD 100 R/M	(DISP-LO),(DISP-HI), DATA-8	AND	REG8/MEM8,IMMED8
80	1000 0000	MOD 101 R/M	(DISP-LO),(DISP-HI), DATA-8	SUB	REG8/MEM8,IMMED8
80	1000 0000	MOD 110 R/M	(DISP-LO),(DISP-HI), DATA-8	XOR	REG8/MEM8,IMMED8
80	1000 0000	MOD 111 R/M	(DISP-LO),(DISP-HI), DATA-8	CMP	REG8/MEM8,IMMED8
81	1000 0001	MOD 000 R/M	(DISP-LO),(DISP-HI), DATA-LO,DATA-HI	ADD	REG16/MEM16,IMMED16
81	1000 0001	MOD 001 R/M	(DISP-LO),(DISP-HI), DATA-LO,DATA-HI	OR	REG16/MEM16,IMMED16
81	1000 0001	MOD 010 R/M	(DISP-LO),(DISP-HI), DATA-LO,DATA-HI	ADC	REG16/MEM16,IMMED16
81	1000 0001	MOD 011 R/M	(DISP-LO),(DISP-HI), DATA-LO,DATA-HI	SBB	REG16/MEM16,IMMED16

Table 4-13. Machine Instruction Decoding Guide (Cont'd.)

1ST BYTE HEX	1ST BYTE BINARY	2ND BYTE	BYTES 3,4,5,6	ASM-86 INSTRUCTION FORMAT	
81	1000 0001	MOD 100 R/M	(DISP-LO),(DISP-HI), DATA-LO,DATA-HI	AND	REG16/MEM16,IMMED16
81	1000 0001	MOD 101 R/M	(DISP-LO),(DISP-HI), DATA-LO,DATA-HI	SUB	REG16/MEM16,IMMED16
81	1000 0001	MOD 110 R/M	(DISP-LO),(DISP-HI), DATA-LO,DATA-HI	XOR	REG16/MEM16.IMMED16
81	1000 0001	MOD 111 R/M	(DISP-LO),(DISP-HI), DATA-LO,DATA-HI	CMP	REG16/MEM16,IMMED16
82	1000 0010	MOD 000 R/M	(DISP-LO),(DISP-HI), DATA-8	ADD	REG8/MEM8,IMMED8
82	1000 0010	MOD 001 R/M		(not used)	
82	1000 0010	MOD 010 R/M	(DISP-LO),(DISP-HI). DATA-8	ADC	REG8/MEM8,IMMED8
82	1000 0010	MOD 011 R/M	(DISP-LO),(DISP-HI), DATA-8	SBB	REG8/MEM8,IMMED8
82	1000 0010	MOD 100 R/M		(not used)	
82	1000 0010	MOD 101 R/M	(DISP-LO),(DISP-HI), DATA-8	SUB	REG8/MEM8,IMMED8
82	1000 0010	MOD 110 R/M		(not used)	
82	1000 0010	MOD 111 R/M	(DISP-LO),(DISP-HI), DATA-8	CMP	REG8/MEM8,IMMED8
83	1000 0011	MOD 000 R/M	(DISP-LO),(DISP-HI), DATA-SX	ADD	REG16/MEM16, IMMED8
83	1000 0011	MOD 001 R/M		(not used)	
83	1000 0011	MOD 010 R/M	(DISP-LO), (DISP-HI), DATA-SX	ADC	REG16/MEM16,IMMED8
83	1000 0011	MOD 011 R/M	(DISP-LO),(DISP-HI), DATA-SX	SBB	REG16/MEM16,IMMED8
83	1000 0011	MOD 100 R/M		(not used)	
83	1000 0011	MOD 101 R/M	(DISP-LO),(DISP-HI), DATA-SX	SUB	REG16/MEM16,IMMED8
83	1000 0011	MOD 110 R/M		(not used)	
83	1000 0011	MOD 111 R/M	(DISP-LO),(DISP-HI), DATA-SX	CMP	REG16/MEM16,IMMED8
84	1000 0100	MOD REG R/M	(DISP-LO),(DISP-HI)	TEST	REG8/MEM8,REG8
85	1000 0101	MOD REG R/M	(DISP-LO),(DISP-HI)	TEST	REG16/MEM16,REG16
86	1000 0110	MOD REG R/M	(DISP-LO),(DISP-HI)	XCHG	REG8,REG8/MEM8
87	1000 0111	MOD REG R/M	(DISP-LO),(DISP-HI)	XCHG	REG16,REG16/MEM16
88	1000 1000	MOD REG R/M	(DISP-LO),(DISP-HI)	MOV	REG8/MEM8,REG8
89	1000 1001	MOD REG R/M	(DISP-LO),(DISP-HI)	MOV	REG16/MEM16/REG16
8A	1000 1010	MOD REG R/M	(DISP-LO),(DISP-HI)	MOV	REG8,REG8/MEM8
8B	1000 1011	MOD REG R/M	(DISP-LO),(DISP-HI)	MOV	REG16,REG16/MEM16
8C	1000 1100	MOD 0SR R/M	(DISP-LO),(DISP-HI)	MOV	REG16/MEM16,SEGREG
8C	1000 1100	MOD 1-- R/M		(not used)	
8D	1000 1101	MOD REG R/M	(DISP-LO),(DISP-HI)	LEA	REG16,MEM16
8E	1000 1110	MOD 0SR R/M	(DISP-LO),(DISP-HI)	MOV	SEGREG,REG16/MEM16
8E	1000 1110	MOD 1-- R/M		(not used)	
8F	1000 1111	MOD 000 R/M	(DISP-LO),(DISP-HI)	POP	REG16/MEM16
8F	1000 1111	MOD 001 R/M		(not used)	
8F	1000 1111	MOD 010 R/M		(not used)	

Table 4-13. Machine Instruction Decoding Guide (Cont'd.)

1ST BYTE HEX	1ST BYTE BINARY	2ND BYTE	BYTES 3,4,5,6	ASM-86 INSTRUCTION FORMAT	
8F	1000 1111	MOD 011 R/M		(not used)	
8F	1000 1111	MOD 100 R/M		(not used)	
8F	1000 1111	MOD 101 R/M		(not used)	
8F	1000 1111	MOD 110 R/M		(not used)	
8F	1000 1111	MOD 111 R/M		(not used)	
90	1001 0000			NOP	(exchange AX,AX)
91	1001 0001			XCHG	AX,CX
92	1001 0010			XCHG	AX,DX
93	1001 0011			XCHG	AX,BX
94	1001 0100			XCHG	AX,SP
95	1001 0101			XCHG	AX,BP
96	1001 0110			XCHG	AX,SI
97	1001 0111			XCHG	AX,DI
98	1001 1000			CBW	
99	1001 1001			CWD	
9A	1001 1010	DISP-LO	DISP-HI,SEG-LO, SEG-HI	CALL	FAR__PROC
9B	1001 1011			WAIT	
9C	1001 1100			PUSHF	
9D	1001 1101			POPF	
9E	1001 1110			SAHF	
9F	1001 1111			LAHF	
A0	1010 0000	ADDR-LO	ADDR-HI	MOV	AL,MEM8
A1	1010 0001	ADDR-LO	ADDR-HI	MOV	AX,MEM16
A2	1010 0010	ADDR-LO	ADDR-HI	MOV	MEM8,AL
A3	1010 0011	ADDR-LO	ADDR-HI	MOV	MEM16,AL
A4	1010 0100			MOVS	DEST-STR8,SRC-STR8
A5	1010 0101			MOVS	DEST-STR16,SRC-STR16
A6	1010 0110			CMPS	DEST-STR8,SRC-STR8
A7	1010 0111			CMPS	DEST-STR16,SRC-STR16
A8	1010 1000	DATA-8		TEST	AL,IMMED8
A9	1010 1001	DATA-LO	DATA-HI	TEST	AX,IMMED16
AA	1010 1010			STOS	DEST-STR8
AB	1010 1011			STOS	DEST-STR16
AC	1010 1100			LODS	SRC-STR8
AD	1010 1101			LODS	SRC-STR16
AE	1010 1110			SCAS	DEST-STR8
AF	1010 1111			SCAS	DEST-STR16
B0	1011 0000	DATA-8		MOV	AL,IMMED8
B1	1011 0001	DATA-8		MOV	CL,IMMED8
B2	1011 0010	DATA-8		MOV	DL,IMMED8
B3	1011 0011	DATA-8		MOV	BL,IMMED8
B4	1011 0100	DATA-8		MOV	AH,IMMED8
B5	1011 0101	DATA-8		MOV	CH,IMMED8
B6	1011 0110	DATA-8		MOV	DH,IMMED8
B7	1011 0111	DATA-8		MOV	BH,IMMED8
B8	1011 1000	DATA-LO	DATA-HI	MOV	AX,IMMED16
B9	1011 1001	DATA-LO	DATA-HI	MOV	CX,IMMED16
BA	1011 1010	DATA-LO	DATA-HI	MOV	DX,IMMED16
BB	1011 1011	DATA-LO	DATA-HI	MOV	BX,IMMED16

Table 4-13. Machine Instruction Decoding Guide (Cont'd.)

1ST BYTE		2ND BYTE	BYTES 3,4,5,6	ASM-86 INSTRUCTION FORMAT	
HEX	BINARY				
BC	1011 1100	DATA-LO	DATA-HI	MOV	SP,IMMED16
BD	1011 1101	DATA-LO	DATA-HI	MOV	BP,IMMED16
BE	1011 1110	DATA-LO	DATA-HI	MOV	SI,IMMED16
BF	1011 1111	DATA-LO	DATA-HI	MOV	DI,IMMED16
C0	1100 0000			(not used)	
C1	1100 0001			(not used)	
C2	1100 0010	DATA-LO	DATA-HI	RET	IMMED16 (intraseg)
C3	1100 0011			RET	(intrasegment)
C4	1100 0100	MOD REG R/M	(DISP-LO),(DISP-HI)	LES	REG16,MEM16
C5	1100 0101	MOD REG R/M	(DISP-LO),(DISP-HI)	LDS	REG16,MEM16
C6	1100 0110	MOD 000 R/M	(DISP-LO),(DISP-HI), DATA-8	MOV	MEM8,IMMED8
C6	1100 0110	MOD 001 R/M		(not used)	
C6	1100 0110	MOD 010 R/M		(not used)	
C6	1100 0110	MOD 011 R/M		(not used)	
C6	1100 0110	MOD 100 R/M		(not used)	
C6	1100 0110	MOD 101 R/M		(not used)	
C6	1100 0110	MOD 110 R/M		(not used)	
C6	1100 0110	MOD 111 R/M		(not used)	
C7	1100 0111	MOD 000 R/M	(DISP-LO),(DISP-HI), DATA-LO,DATA-HI	MOV	MEM16,IMMED16
C7	1100 0111	MOD 001 R/M		(not used)	
C7	1100 0111	MOD 010 R/M		(not used)	
C7	1100 0111	MOD 011 R/M		(not used)	
C7	1100 0111	MOD 100 R/M		(not used)	
C7	1100 0111	MOD 101 R/M		(not used)	
C7	1100 0111	MOD 110 R/M		(not used)	
C7	1100 0111	MOD 111 R/M		(not used	
C8	1100 1000			(not used)	
C9	1100 1001			(not used)	
CA	1100 1010	DATA-LO	DATA-HI	RET	IMMED16 (intersegment)
CB	1100 1011			RET	(intersegment)
CC	1100 1100			INT	3
CD	1100 1101	DATA-8		INT	IMMED8
CE	1100 1110			INTO	
CF	1100 1111			IRET	
D0	1101 0000	MOD 000 R/M	(DISP-LO),(DISP-HI)	ROL	REG8/MEM8,1
D0	1101 0000	MOD 001 R/M	(DISP-LO),(DISP-HI)	ROR	REG8/MEM8,1
D0	1101 0000	MOD 010 R/M	(DISP-LO),(DISP-HI)	RCL	REG8/MEM8,1
D0	1101 0000	MOD 011 R/M	(DISP-LO),(DISP-HI)	RCR	REG8/MEM8,1
D0	1101 0000	MOD 100 R/M	(DISP-LO),(DISP-HI)	SAL/SHL	REG8/MEM8,1
D0	1101 0000	MOD 101 R/M	(DISP-LO),(DISP-HI)	SHR	REG8/MEM8,1
D0	1101 0000	MOD 110 R/M		(not used)	
D0	1101 0000	MOD 111 R/M	(DISP-LO),(DISP-HI)	SAR	REG8/MEM8,1
D1	1101 0001	MOD 000 R/M	(DISP-LO),(DISP-HI)	ROL	REG16/MEM16,1
D1	1101 0001	MOD 001 R/M	(DISP-LO),(DISP-HI)	ROR	REG16/MEM16,1
D1	1101 0001	MOD 010 R/M	(DISP-LO),(DISP-HI)	RCL	REG16/MEM16,1
D1	1101 0001	MOD 011 R/M	(DISP-LO),(DISP-HI)	RCR	REG16/MEM16,1
D1	1101 0001	MOD 100 R/M	(DISP-LO),(DISP-HI)	SAL/SHL	REG16/MEM16,1

Table 4-13. Machine Instruction Decoding Guide (Cont'd.)

1ST BYTE HEX	1ST BYTE BINARY	2ND BYTE	BYTES 3,4,5,6	ASM-86 INSTRUCTION FORMAT	
D1	1101 0001	MOD 101 R/M	(DISP-LO),(DISP-HI)	SHR	REG16/MEM16,1
D1	1101 0001	MOD 110 R/M		(not used)	
D1	1101 0001	MOD 111 R/M	(DISP-LO),(DISP-HI)	SAR	REG16/MEM16,1
D2	1101 0010	MOD 000 R/M	(DISP-LO),(DISP-HI)	ROL	REG8/MEM8,CL
D2	1101 0010	MOD 001 R/M	(DISP-LO),(DISP-HI)	ROR	REG8/MEM8,CL
D2	1101 0010	MOD 010 R/M	(DISP-LO),(DISP-HI)	RCL	REG8/MEM8,CL
D2	1101 0010	MOD 011 R/M	(DISP-LO),(DISP-HI)	RCR	REG8/MEM8,CL
D2	1101 0010	MOD 100 R/M	(DISP-LO),(DISP-HI)	SAL/SHL	REG8/MEM8,CL
D2	1101 0010	MOD 101 R/M	(DISP-LO),(DISP-HI)	SHR	REG8/MEM8,CL
D2	1101 0010	MOD 110 R/M		(not used)	
D2	1101 0010	MOD 111 R/M	(DISP-LO),(DISP-HI)	SAR	REG8/MEM8,CL
D3	1101 0011	MOD 000 R/M	(DISP-LO),(DISP-HI)	ROL	REG16/MEM16,CL
D3	1101 0011	MOD 001 R/M	(DISP-LO),(DISP-HI)	ROR	REG16/MEM16,CL
D3	1101 0011	MOD 010 R/M	(DISP-LO),(DISP-HI)	RCL	REG16/MEM16,CL
D3	1101 0011	MOD 011 R/M	(DISP-LO),(DISP-HI)	RCR	REG16/MEM16,CL
D3	1101 0011	MOD 100 R/M	(DISP-LO),(DISP-HI)	SAL/SHL	REG16/MEM16,CL
D3	1101 0011	MOD 101 R/M	(DISP-LO),(DISP-HI)	SHR	REG16/MEM16,CL
D3	1101 0011	MOD 110 R/M		(not used)	
D3	1101 0011	MOD 111 R/M	(DISP-LO),(DISP-HI)	SAR	REG16/MEM16,CL
D4	1101 0100	00001010		AAM	
D5	1101 0101	00001010		AAD	
D6	1101 0110			(not used)	
D7	1101 0111			XLAT	SOURCE-TABLE
D8	1101 1000	MOD 000 R/M			
	1XXX	MOD YYY R/M	(DISP-LO), (DISP-HI)	ESC	OPCODE,SOURCE
DF	1101 1111	MOD 111 R/M			
E0	1110 0000	IP-INC-8		LOOPNE/ LOOPNZ	SHORT-LABEL
E1	1110 0001	IP-INC-8		LOOPE/ LOOPZ	SHORT-LABEL
E2	1110 0010	IP-INC-8		LOOP	SHORT-LABEL
E3	1110 0011	IP-INC-8		JCXZ	SHORT-LABEL
E4	1110 0100	DATA-8		IN	AL,IMMED8
E5	1110 0101	DATA-8		IN	AX,IMMED8
E6	1110 0110	DATA-8		OUT	AL,IMMED8
E7	1110 0111	DATA-8		OUT	AX,IMMED8
E8	1110 1000	IP-INC-LO	IP-INC-HI	CALL	NEAR-PROC
E9	1110 1001	IP-INC-LO	IP-INC-HI	JMP	NEAR-LABEL
EA	1110 1010	IP-LO	IP-HI,CS-LO,CS-HI	JMP	FAR-LABEL
EB	1110 1011	IP-INC8		JMP	SHORT-LABEL
EC	1110 1100			IN	AL,DX
ED	1110 1101			IN	AX,DX
EE	1110 1110			OUT	AL,DX
EF	1110 1111			OUT	AX,DX
F0	1111 0000			LOCK	(prefix)
F1	1111 0001			(not used)	
F2	1111 0010			REPNE/REPNZ	
F3	1111 0011			REP/REPE/REPZ	
F4	1111 0100			HLT	
F5	1111 0101			CMC	

Mnemonics © Intel, 1978

Table 4-13. Machine Instruction Decoding Guide (Cont'd.)

HEX	BINARY	2ND BYTE	BYTES 3,4,5,6	ASM-86 INSTRUCTION FORMAT	
F6	1111 0110	MOD 000 R/M	(DISP-LO),(DISP-HI), DATA-8	TEST	REG8/MEM8,IMMED8
F6	1111 0110	MOD 001 R/M		(not used)	
F6	1111 0110	MOD 010 R/M	(DISP-LO),(DISP-HI)	NOT	REG8/MEM8
F6	1111 0110	MOD 011 R/M	(DISP-LO),(DISP-HI)	NEG	REG8/MEM8
F6	1111 0110	MOD 100 R/M	(DISP-LO),(DISP-HI)	MUL	REG8/MEM8
F6	1111 0110	MOD 101 R/M	(DISP-LO),(DISP-HI)	IMUL	REG8/MEM8
F6	1111 0110	MOD 110 R/M	(DISP-LO),(DISP-HI)	DIV	REG8/MEM8
F6	1111 0110	MOD 111 R/M	(DISP-LO),(DISP-HI)	IDIV	REG8/MEM8
F7	1111 0111	MOD 000 R/M	(DISP-LO),(DISP-HI), DATA-LO,DATA-HI	TEST	REG16/MEM16,IMMED16
F7	1111 0111	MOD 001 R/M		(not used)	
F7	1111 0111	MOD 010 R/M	(DISP-LO),(DISP-HI)	NOT	REG16/MEM16
F7	1111 0111	MOD 011 R/M	(DISP-LO),(DISP-HI)	NEG	REG16/MEM16
F7	1111 0111	MOD 100 R/M	(DISP-LO),(DISP-HI)	MUL	REG16/MEM16
F7	1111 0111	MOD 101 R/M	(DISP-LO),(DISP-HI)	IMUL	REG16/MEM16
F7	1111 0111	MOD 110 R/M	(DISP-LO),(DISP-HI)	DIV	REG16/MEM16
F7	1111 0111	MOD 111 R/M	(DISP-LO),(DISP-HI)	IDIV	REG16/MEM16
F8	1111 1000			CLC	
F9	1111 1001			STC	
FA	1111 1010			CLI	
FB	1111 1011			STI	
FC	1111 1100			CLD	
FD	1111 1101			STD	
FE	1111 1110	MOD 000 R/M	(DISP-LO),(DISP-HI)	INC	REG8/MEM8
FE	1111 1110	MOD 001 R/M	(DISP-LO),(DISP-HI)	DEC	REG8/MEM8
FE	1111 1110	MOD 010 R/M		(not used)	
FE	1111 1110	MOD 011 R/M		(not used)	
FE	1111 1110	MOD 100 R/M		(not used)	
FE	1111 1110	MOD 101 R/M		(not used)	
FE	1111 1110	MOD 110 R/M		(not used)	
FE	1111 1110	MOD 111 R/M		(not used)	
FF	1111 1111	MOD 000 R/M	(DISP-LO),(DISP-HI)	INC	MEM16
FF	1111 1111	MOD 001 R/M	(DISP-LO),(DISP-HI)	DEC	MEM16
FF	1111 1111	MOD 010 R/M	(DISP-LO),(DISP-HI)	CALL	REG16/MEM16 (intra)
FF	1111 1111	MOD 011 R/M	(DISP-LO),(DISP-HI)	CALL	MEM16 (intersegment)
FF	1111 1111	MOD 100 R/M	(DISP-LO),(DISP-HI)	JMP	REG16/MEM16 (intra)
FF	1111 1111	MOD 101 R/M	(DISP-LO),(DISP-HI)	JMP	MEM16 (intersegment)
FF	1111 1111	MOD 110 R/M	(DISP-LO),(DISP-HI)	PUSH	MEM16
FF	1111 1111	MOD 111 R/M		(not used)	

Appendix I
8253 Programmable interval timer data sheets

The IBM-PC has one of these for internal timing functions. Timer 0 is used for the TimeOfDay functions and generates interrupts on the IRQ0 line every 55 ms. Timer 2 is used for sending tones to the speaker. Timer 1 is used internally for refreshing the dynamic RAM memory chips.

■ **MCS-85™ Compatible 8253-5**

■ **3 Independent 16-Bit Counters**

■ **DC to 2.6 MHz**

■ **Programmable Counter Modes**

■ **Count Binary or BCD**

■ **Single +5V Supply**

■ **Available in EXPRESS**
— **Standard Temperature Range**
— **Extended Temperature Range**

The Intel® 8253 is a programmable counter/timer device designed for use as an Intel microcomputer peripheral. It uses NMOS technology with a single +5V supply and is packaged in a 24-pin plastic DIP.

It is organized as 3 independent 16-bit counters, each with a count rate of up to 2.6 MHz. All modes of operation are software programmable.

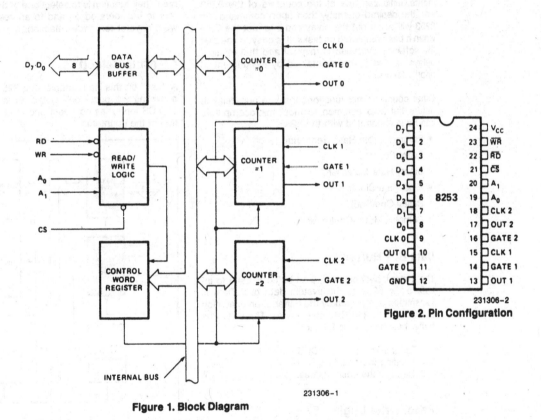

231306-1

Figure 1. Block Diagram

231306-2

Figure 2. Pin Configuration

FUNCTIONAL DESCRIPTION

General

The 8253 is programmable interval timer/counter specifically designed for use with the Intel™ Micro-computer systems. Its function is that of a general purpose, multi-timing element that can be treated as an array of I/O ports in the system software.

The 8253 solves one of the most common problems in any microcomputer system, the generation of ac-curate time delays under software control. Instead of setting up timing loops in systems software, the pro-grammer configures the 8253 to match his require-ments, initializes one of the counters of the 8253 with the desired quantity, then upon command the 8253 will count out the delay and interrupt the CPU when it has completed its tasks. It is easy to see that the software overhead is minimal and that multiple delays can easily be maintained by assignment of priority levels.

Other counter/timer functions that are non-delay in nature but also common to most microcomputers can be implemented with the 8253.

- Programmable Rate Generator
- Event Counter
- Binary Rate Multiplier
- Real Time Clock
- Digital One-Shot
- Complex Motor Controller

Data Bus Buffer

The 3-state, bi-directional, 8-bit buffer is used to in-terface the 8253 to the system data bus. Data is transmitted or received by the buffer upon execution of INput or OUTput CPU instructions. The Data Bus Buffer has three basic functions.

1. Programming the MODES of the 8253.
2. Loading the count registers.
3. Reading the count values.

Read/Write Logic

The Read/Write Logic accepts inputs from the sys-tem bus and in turn generates control signals for overall device operation. It is enabled or disabled by CS so that no operation can occur to change the function unless the device has been selected by the system logic.

$\overline{RD}$ (Read)

A "low" on this input informs the 8253 that the CPU is inputting data in the form of a counters value.

$\overline{WR}$ (Write)

A "low" on this input informs the 8253 that the CPU is outputting data in the form of mode information or loading counters.

A0, A1

These inputs are normally connected to the address bus. Their function is to select one of the three coun-ters to be operated on and to address the control word register for mode selection.

$\overline{CS}$ (Chip Select)

A "low" on this input enables the 8253. No reading or writing will occur unless the device is selected. The $\overline{CS}$ input has no effect upon the actual opera-tion of the counters.

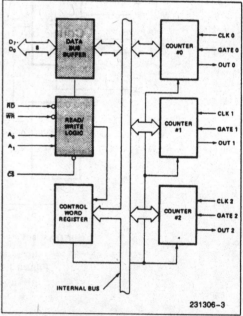

Figure 3. Block Diagram Showing Data Bus Buffer and Read/Write Logic Functions

$\overline{CS}$	$\overline{RD}$	$\overline{WR}$	A_1	A_0	
0	1	0	0	0	Load Counter No. 0
0	1	0	0	1	Load Counter No. 1
0	1	0	1	0	Load Counter No. 2
0	1	0	1	1	Write Mode Word
0	0	1	0	0	Read Counter No. 0
0	0	1	0	1	Read Counter No. 1
0	0	1	1	0	Read Counter No. 2
0	0	1	1	1	No-Operation 3-State
1	X	X	X	X	Disable 3-State
0	1	1	X	X	No-Operation 3-State

Control Word Register

The Control Word Register is selected when A0, A1 are 11. It then accepts information from the data bus buffer and stores it in a register. The information stored in this register controls the operation MODE of each counter, selection of binary or BCD counting and the loading of each count register.

The Control Word Register can only be written into; no read operation of its contents is available.

Counter #0, Counter #1, Counter #2

These three functional blocks are identical in operation so only a single counter will be described. Each Counter consists of a single, 16-bit, pre-settable, DOWN counter. The counter can operate in either binary or BCD and its input, gate and output are configured by the selection of MODES stored in the Control Word Register.

The counters are fully independent and each can have separate MODE configuration and counting operation, binary or BCD. Also, there are special features in the control word that handle the loading of the count value so that software overhead can be minimized for these functions.

The reading of the contents of each counter is available to the programmer with simple READ operations for event counting applications and special commands and logic are included in the 8253 so that the contents of each counter can be read "on the fly" without having to inhibit the clock input.

8253 SYSTEM INTERFACE

The 8253 is a component of the Intel™ Microcomputer systems and interfaces in the same manner as all other peripherals of the family. It is treated by the systems software as an array of peripheral I/O ports; three are counters and the fourth is a control register for MODE programming.

Basically, the select inputs A0, A1 connect to the A0, A1 address bus signals of the CPU. The $\overline{CS}$ can be derived directly from the address bus using a linear select method. Or it can be connected to the output of a decoder, such as an Intel 8205 for larger systems.

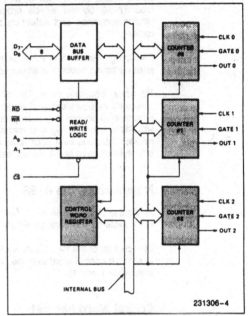

Figure 4. Block Diagram Showing Control Word Register and Counter Functions

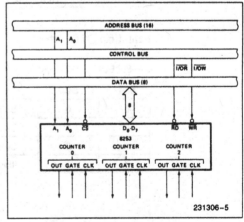

Figure 5. 8253 System Interface

OPERATIONAL DESCRIPTION

General

The complete functional definition of the 8253 is programmed by the systems software. A set of control words *must* be sent out by the CPU to initialize each counter of the 8253 with the desired MODE and quantity information. Prior to initialization, the MODE, count, and output of all counters is undefined. These control words program the MODE, Loading sequence and selection of binary or BCD counting.

Once programmed, the 8253 is ready to perform whatever timing tasks it is assigned to accomplish.

The actual counting operation of each counter is completely independent and additional logic is provided on-chip so that the usual problems associated with efficient monitoring and management of external, asynchronous events or rates to the microcomputer system have been eliminated.

Programming the 8253

All of the MODES for each counter are programmed by the systems software by simple I/O operations.

Each counter of the 8253 is individually programmed by writing a control word into the Control Word Register. (A0, A1 = 11)

Control Word Format

D_7	D_6	D_5	D_4	D_3	D_2	D_1	D_0
SC1	SC0	RL1	RL0	M2	M1	M0	BCD

Definition Of Control

SC—SELECT COUNTER:

SC1	SC0	
0	0	Select Counter 0
0	1	Select Counter 1
1	0	Select Counter 2
1	1	Illegal

RL—READ/LOAD:

RL1	RL0	
0	0	Counter Latching operation (see READ/WRITE Procedure Section).
1	0	Read/Load most significant byte only.
0	1	Read/Load least significant byte only.
1	1	Read/Load least significant byte first, then most significant byte.

M—MODE:

M2	M1	M0	
0	0	0	Mode 0
0	0	1	Mode 1
X	1	0	Mode 2
X	1	1	Mode 3
1	0	0	Mode 4
1	0	1	Mode 5

BCD:

0	Binary Counter 16-Bits
1	Binary Coded Decimal (BCD) Counter (4 Decades)

Counter Loading

The count register is not loaded until the count value is written (one or two bytes, depending on the mode selected by the RL bits), followed by a rising edge and a falling edge of the clock. Any read of the counter prior to that falling clock edge may yield invalid data.

MODE DEFINITION

MODE 0: Interrupt on Terminal Count. The output will be initially low after the mode set operation. After the count is loaded into the selected count register, the output will remain low and the counter will count. When terminal count is reached, the output will go high and remain high until the selected count register is reloaded with the mode or a new count is loaded. The counter continues to decrement after terminal count has been reached.

Rewriting a counter register during counting results in the following:

(1) Write 1st byte stops the current counting.
(2) Write 2nd byte starts the new count.

MODE 1: Programmable One-Shot. The output will go low on the count following the rising edge of the gate input.

The output will go high on the terminal count. If a new count value is loaded while the output is low it will not affect the duration of the one-shot pulse until the succeeding trigger. The current count can be read at any time without affecting the one-shot pulse.

The one-shot is retriggerable, hence the output will remain low for the full count after any rising edge of the gate input.

MODE 2: Rate Generator. Divide by N counter. The output will be low for one period of the input clock. The period from one output pulse to the next equals the number of input counts in the count register. If the count register is reloaded between output pulses the present period will not be affected, but the subsequent period will reflect the new value.

The gate input, when low, will force the output high. When the gate input goes high, the counter will start from the initial count. Thus, the gate input can be used to synchronize the counter.

When this mode is set, the output will remain high until after the count register is loaded. The output then can also be synchronized by software.

MODE 3: Square Wave Rate Generator. Similar to MODE 2 except that the output will remain high until one half the count has been completed (or even numbers) and go low for the other half of the count. This is accomplished by decrementing the counter by two on the falling edge of each clock pulse. When the counter reaches terminal count, the state of the output is changed and the counter is reloaded with the full count and the whole process is repeated.

If the count is odd and the output is high, the first clock pulse (after the count is loaded) decrements the count by 1. Subsequent clock pulses decrement the clock by 2. After timeout, the output goes low and the full count is reloaded. The first clock pulse (following the reload) decrements the counter by 3. Subsequent clock pulses decrement the count by 2 until timeout. Then the whole process is repeated. In this way, if the count is odd, the output will be high for $(N + 1)/2$ counts and low for $(N - 1)/2$ counts.

In Modes 2 and 3, if a CLK source other than the system clock is used, GATE should be pulsed immediately following $\overline{WR}$ of a new count value.

MODE 4: Software Triggered Strobe. After the mode is set, the output will be high. When the count is loaded, the counter will begin counting. On terminal count, the output will go low for one input clock period, then will go high again.

If the count register is reloaded during counting, the new count will be loaded on the next CLK pulse. The count will be inhibited while the GATE input is low.

MODE 5: Hardware Triggered Strobe. The counter will start counting after the rising edge of the trigger input and will go low for one clock period when the terminal count is reached. The counter is retriggerable. The output will not go low until the full count after the rising edge of any trigger.

Signal Status Modes	Low Or Going Low	Rising	High
0	Disables counting	—	Enables counting
1	—	1) Initiates counting 2) Resets output after next clock	—
2	1) Disables counting 2) Sets output immediately high	1) Reloads counter 2) Initiates counting	Enables counting
3	1) Disables counting 2) Sets output immediately high	1) Reloads counter 2) Initiates counting	Enables counting
4	Disables counting	—	Enables counting
5	—	Initiates counting	—

Figure 6. Gate Pin Operations Summary

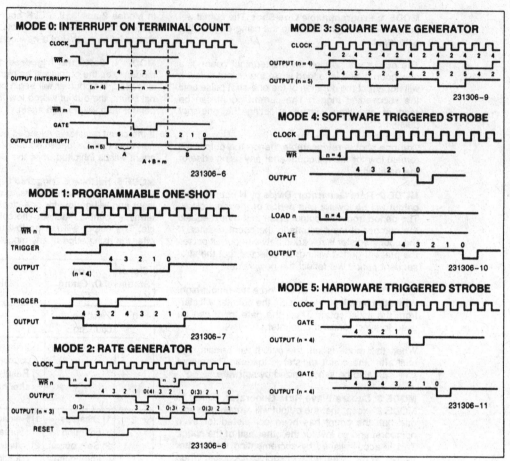

Figure 7. 8253 Timing Diagrams

8253 READ/WRITE PROCEDURE

Write Operations

The systems software must program each counter of the 8253 with the mode and quantity desired. The programmer must write out to the 8253 a MODE control word and the programmed number of count register bytes (1 or 2) prior to actually using the selected counter.

The actual order of the programming is quite flexible. Writing out of the MODE control word can be in any sequence of counter selection, e.g., counter #0 does not have to be first or counter #2 last. Each counter's MODE control word register has a separate address so that its loading is completely sequence independent. (SC0, SC1).

The loading of the Count Register with the actual count value, however, must be done in exactly the sequence programmed in the MODE control word (RL0, RL1). This loading of the counter's count register is still sequence independent like the MODE control word loading, but when a selected count register is to be loaded it *must* be loaded with the number of bytes programmed in the MODE control word (RL0, RL1). The one or two bytes to be loaded in the count register do not have to follow the associated MODE control word. They can be programmed at any time following the MODE control word loading as long as the correct number of bytes is loaded in order.

All counters are down counters. Thus, the value loaded into the count register will actually be decremented. Loading all zeros into a count register will result in the maximum count (2^{16} for Binary or 10^4 for BCD). In MODE 0 the new count will not restart until the load has been completed. It will accept one of two bytes depending on how the MODE control words (RL0, RL1) are programmed. Then proceed with the restart operation.

	MODE Control Word Counter n
LSB	Counter Register byte Counter n
MSB	Counter Register byte Counter n

NOTE:
Format shown is a simple example of loading the 8253 and does not imply that it is the only format that can be used.

Figure 8. Programming Format

			A1	A0
No. 1		MODE Control Word Counter 0	1	1
No. 2		MODE Control Word Counter 1	1	1
No. 3		MODE Control Word Counter 2	1	1
No. 4	LSB	Count Register Byte Counter 1	0	1
No. 5	MSB	Count Register Byte Counter 1	0	1
No. 6	LSB	Count Register Byte Counter 2	1	0
No. 7	MSB	Count Register Byte Counter 2	1	0
No. 8	LSB	Count Register Byte Counter 0	0	0
No. 9	MSB	Count Register Byte Counter 0	0	0

NOTE:
The exclusive addresses of each counter's count register make the task of programming the 8253 a very simple matter, and maximum effective use of the device will result if this feature is fully initilized.

Figure 9. Alternate Programming Formats

Read Operations

In most counter applications it becomes necessary to read the value of the count in progress and make a computational decision based on this quantity. Event counters are probably the most common application that uses this function. The 8253 contains logic that will allow the programmer to easily read the contents of any of the three counters without disturbing the actual count in progress.

There are two methods that the programmer can use to read the value of the counters. The first method involves the use of simple I/O read operations of the selected counter. By controlling the A0, A1 inputs to the 8253 the programmer can select the counter to be read (remember that no read operation of the mode register is allowed A0, A1-11). The only requirement with this method is that in order to assure a stable count reading the actual operation of the selected counter *must be inhibited* either by controlling the Gate input or by external logic that inhibits the clock input. The contents of the counter selected will be available as follows:

First I/O Read contains the least significant byte (LSB).

Second I/O Read contains the most significant byte (MSB).

Due to the internal logic of the 8253 it is absolutely necessary to complete the entire reading procedure. If two bytes are programmed to be read, then two bytes *must* be read before any loading WR command can be sent to the same counter.

Read Operation Chart

A1	A0	RD	
0	0	0	Read Counter No. 0
0	1	0	Read Counter No. 1
1	0	0	Read Counter No. 2
1	1	0	Illegal

Reading While Counting

In order for the programmer to read the contents of any counter without effecting or disturbing the counting operation the 8253 has special internal logic that can be accessed using simple $\overline{WR}$ commands to the MODE register. Basically, when the programmer wishes to read the contents of a selected counter "on the fly" he loads the MODE register with a special code which latches the present count value into a storage register so that its contents contain an accurate, stable quantity. The programmer then issues a normal read command to the selected counter and the contents of the latched register is available.

MODE Register for Latching Count

A0, A1 = 11

D7	D6	D5	D4	D3	D2	D1	D0
SC1	SC0	0	0	X	X	X	X

SC1, SC0— specify counter to be latched.
D5, D4   — 00 designates counter latching operation.
X   — don't care.

The same limitation applies to this mode of reading the counter as the previous method. That is, it is mandatory to complete the entire read operation as programmed. This command has no effect on the counter's mode.

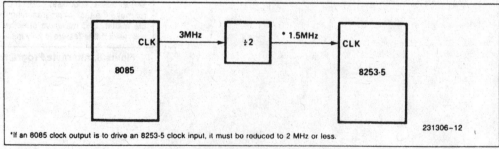

*If an 8085 clock output is to drive an 8253-5 clock input, it must be reduced to 2 MHz or less.

Figure 10. MCS-85™ Clock Interface*

ABSOLUTE MAXIMUM RATINGS*

Ambient Temperature Under Bias ......0°C to 70°C

Storage Temperature .... −65°C to +150°C

Voltage On Any Pin
with Respect to Ground............ −0.5V to 7V

Power Dissipation ....1 Watt

*Notice: Stresses above those listed under "Absolute Maximum Ratings" may cause permanent damage to the device. This is a stress rating only and functional operation of the device at these or any other conditions above those indicated in the operational sections of this specification is not implied. Exposure to absolute maximum rating conditions for extended periods may affect device reliability.

D.C. CHARACTERISTICS T_A = 0°C to 70°C, V_{CC} = 5V ±10%*

Symbol	Parameter	Min	Max	Unit	Test Conditions
V_{IL}	Input Low Voltage	−0.5	0.8	V	
V_{IH}	Input High Voltage	2.2	V_{CC} + .5V	V	
V_{OL}	Output Low Voltage		0.45	V	(Note 1)
V_{OH}	Output High Voltage	2.4		V	(Note 2)
I_{IL}	Input Load Current		±10	μA	V_{IN} = V_{CC} to 0V
I_{OFL}	Output Float Leakage		±10	μA	V_{OUT} = V_{CC} to 0.45V
I_{CC}	V_{CC} Supply Current		140	mA	

CAPACITANCE T_A = 25°C, V_{CC} = GND = 0V

Symbol	Parameter	Min	Typ	Max	Unit	Test Conditions
C_{IN}	Input Capacitance			10	pF	fc = 1 MHz
$C_{I/O}$	I/O Capacitance			20	pF	Unmeasured pins returned to V_{SS}

A.C. CHARACTERISTICS T_A = 0°C to 70°C, V_{CC} = 5.0V ±10%, GND = 0V*

Bus Parameters[3]

READ CYCLE

Symbol	Parameter	8253		8253-5		Unit
		Min	Max	Min	Max	
t_{AR}	Address Stable before READ	50		30		ns
t_{RA}	Address Hold Time for READ	5		5		ns
t_{RR}	READ Pulse Width	400		300		ns
t_{RD}	Data Delay from READ[4]		300		200	ns
t_{DF}	READ to Data Floating	25	125	25	100	ns
t_{RV}	Recovery Time between READ and Any Other Control Signal	1		1		μs

A.C. CHARACTERISTICS (Continued)

WRITE CYCLE

Symbol	Parameter	8253		8253-5		Unit
		Min	Max	Min	Max	
t_{AW}	Address Stable before $\overline{WRITE}$	50		30		ns
t_{WA}	Address Hold Time for $\overline{WRITE}$	30		30		ns
t_{WW}	$\overline{WRITE}$ Pulse Width	400		300		ns
t_{DW}	Data Set Up Time for $\overline{WRITE}$	300		250		ns
t_{WD}	Data Hold Time for $\overline{WRITE}$	40		30		ns
t_{RV}	Recovery Time between $\overline{WRITE}$ and Any Other Control Signal	1		1		μs

CLOCK AND GATE TIMING

Symbol	Parameter	8253		8253-5		Unit
		Min	Max	Min	Max	
t_{CLK}	Clock Period	380	dc	380	dc	ns
t_{PWH}	High Pulse Width	230		230		ns
t_{PWL}	Low Pulse Width	150		150		ns
t_{GW}	Gate Width High	150		150		ns
t_{GL}	Gate Width Low	100		100		ns
t_{GS}	Gate Set Up Time to CLK ↑	100		100		ns
t_{GH}	Gate Hold Time after CLK ↑	50		50		ns
t_{OD}	Output Delay from CLK ↓ (4)		400		400	ns
t_{ODG}	Output Delay from Gate ↓ (4)		300		300	ns

NOTES:
1. $I_{OL} = 2.2$ mA.
2. $I_{OH} = -400$ μA.
3. AC timings measured at V_{OH} 2.2, $V_{OL} = 0.8$.
4. $C_L = 150$ pF.
*For Extended Temperature EXPRESS, use M8253 electrical parameters.

A.C. TESTING INPUT, OUTPUT WAVEFORM

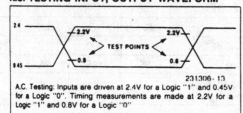

231306-13
A.C. Testing: Inputs are driven at 2.4V for a Logic "1" and 0.45V for a Logic "0". Timing measurements are made at 2.2V for a Logic "1" and 0.8V for a Logic "0"

A.C. TESTING LOAD CIRCUIT

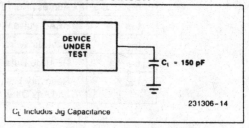

231306-14

C_L Includes Jig Capacitance

WAVEFORMS

WRITE TIMING

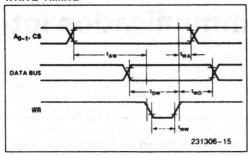

231306-15

READ TIMING

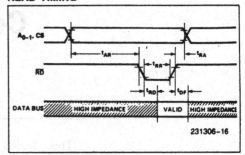

231306-16

CLOCK AND GATE TIMING

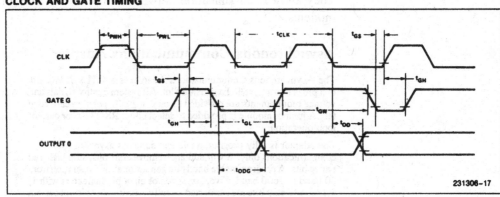

231306-17

Appendix J
8250/8251 Programmable Communication Interface

This device is commonly called a UART (Universal Asynchronous Receiver/Transmitter). There is one for each COM port of the computer. They allow communication between computers directly or via telephone modems.

Asynchronous Communications Adapter

The Asynchronous Communications Adapter is a 4"H x 5"W card that plugs into a System Expansion Slot. All system control signals and voltage requirements are provided through a 2 x 31 position card edge tab. A jumper module is provided to select either RS-232-C or current loop operation.

The adapter is fully programmable and supports asynchronous communications only. It will add and remove start bits, stop bits, and parity bits. A programmable baud rate generator allows operation from 50 baud to 9600 baud. Five, six, seven or eight bit characters with 1, 1-1/2, or 2 stop bits are supported. A fully prioritized interrupt system controls transmit, receive, error, line status and data set interrupts. Diagnostic capabilities provide loopback functions of transmit/receive and input/output signals.

Figure (22) is a block diagram of the Asynchronous Communications Adapter.

The heart of the adapter is a INS8250 LSI chip or functional equivalent. The following is a summary of the 8250's key features:

- Adds or Delete Standard Asynchronous Communication Bits (Start, Stop, and Parity) to or from Serial Data Stream.

- Full Double Buffering Eliminates Need for Precise Synchronization.

- Independently Controlled Transmit, Receive, Line Status, and Data Set Interrupts.

- Programmable Baud Rate Generator Allows Division of Any Input Clock by 1 to (2^{16}-1) and Generates the Internal 16x Clock.

- Independent Receiver Clock Input.

- MODEM Control Functions Clear to Send (CTS), Request to Send (RTS), Data Set Ready (DSR), Data Terminal Ready (DTR), Ring Indicator (RI), and Carrier Detect.

- Fully Programmable Serial-Interface Characteristics

    - 5-, 6-, 7-, or 8-Bit Characters

    - Even, Odd, or No-Parity Bit Generation and Detection

    - 1-, 1 1/2-, or 2-Stop Bit Generation

    - Baud Rate Generation (DC to 9600 Baud)

- False Start Bit Detection.
- Complete Status Reporting Capabilities.
- Line Break Generation and Detection.
- Internal Diagnostic Capabilities.
    - Loopback Controls for Communications Link Fault Isolation.
    - Break, Parity, Overrun, Framing Error Simulation.
- Full Prioritized Interrupt System Controls.

All communications protocol is a function of the system microcode and must be loaded before the adapter is operational. All pacing of the interface and control signal status must be handled by the system software.

Asynchronous Communications Block Diagram

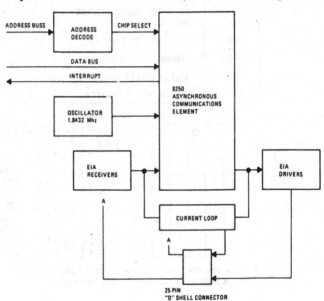

Figure 22. ASYNCHRONOUS COMMUNICATIONS ADAPTER
BLOCK DIAGRAM

Modes of Operation

The different modes of operation are selected by programming the 8250 Asynchronous Communications Element. This is done by selecting the I/O address (3F8 to 3FF) and writing data out to the card. Address bit A0, A1 and A2 select the different registers which define the modes of operation. Also, the Divisor Latch Access Bit (Bit 7) of the line control register is used to select certain registers.

I/O Decode for Communications Adapter

Table 21. I/O Decodes (3F8 to 3FF)

I/O DECODE	REGISTER SELECTED	DLAB STATE	
3F8	TX BUFFER	DLAB=0	(WRITE)
3F8	RX BUFFER	DLAB=0	(READ)
3F8	DIVISOR LATCH LSB	DLAB=1	
3F9	DIVISOR LATCH MSB	DLAB=1	
3F9	INTERRUPT ENABLE REGISTER	DLAB=0	
3FA	INTERRUPT IDENTIFICATION REGISTERS		
3FB	LINE CONTROL REGISTER		
3FC	MODEM CONTROL REGISTER		
3FD	LINE STATUS REGISTER		
3FE	MODEM STATUS REGISTER		

ADDRESS BITS

	A9	A8	A7	A6	A5	A4	A3	A2	A1	A0	DLAB	REGISTER
3F8 to 3FF	1	1	1	1	1	1	1	X	X	X		
								0	0	0	0	Receive Buffer (read), Transmit Holding Reg. (write)
								0	0	1	0	Interrupt Enable
								0	1	0	X	Interrupt Identification
								0	1	1	X	Line Control
								1	0	0	X	Modem Control
								1	0	1	X	Line Status
								1	1	0	X	Modem Status
								1	1	1	X	None
								0	0	0	1	Divisor Latch (LSB)
								0	0	1	1	Divisor Latch (MSB)

A2, A1 and A0 bits are "Don't Cares" and are used to select the different register of the communications chip.

Interrupts

One interrupt line is provided to the system. This interrupt is IRQ4 and will be positive active. To allow the communications card to send interrupts to the system, Bit 3 of the Modem Control Register must be set = 0 (low). At this point, any interrupts allowed by the Interrupt Enable Register will cause an interrupt.

The data format will be as follows:

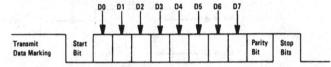

Data Bit 0 is the first bit to be transmitted or received. The adapter automatically inserts the start bit, the correct parity bit if programmed to do so, and the stop bit (1, 1-1/2 or 2 depending on the command in the Line Control Register).

Interface Description

The communications adapter provides an EIA RS-232-C like interface. One 25 pin "D" shell, male type connector is provided to attach various peripheral devices. In addition, a current loop interface is also located in this same connector. A jumper block is provided to manually select either the voltage interface, or the current loop interface.

The current loop interface is provided to attach certain printers provided by IBM Corporation that use this particular type of interface.

Pin 18 + receive current loop data (20Ma)
Pin 25 - receive current loop return (20Ma)
Pin 9 + transmit current loop return (20Ma)
Pin 11 - transmit current loop data (20Ma)

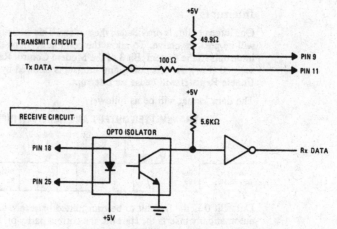

Figure 23. CURRENT LOOP INTERFACE

The voltage interface is a serial interface. It supports certain data and control signals as listed below.

Pin	Signal
Pin 2	Transmit Data
Pin 3	Receive Data
Pin 4	Request to Send
Pin 5	Clear to Send
Pin 6	Data Set Ready
Pin 7	Signal Ground
Pin 8	Carrier Detect
Pin 20	Data Terminal Ready
Pin 22	Ring Indicate

The adapter converts these signals to/from TTL levels to EIA voltage levels. These signals are sampled or generated by the communication control chip. These signals can then be sensed by the system software to determine the state of the interface or peripheral device.

Voltage Interchange Information

Interchange Voltage	Binary State	Signal Condition	Interface Control Function
Positive Voltage =	Binary (0)	= Spacing	=On
Negative Voltage =	Binary (1)	= Marking	=Off

```
                  Invalid Levels
+15V — — — — — — — — — —
                  On Function
 +3V — — — — — — — — —
 0V               Invalid Levels
 -3V — — — — — — — — —
                  Off Function
-15V — — — — — — — —
                  Invalid Levels
```

The signal will be considered in the "marking" condition when the voltage on the interchange circuit, measured at the interface point, is more negative than minus three volts with respect to signal ground. The signal will be considered in the "spacing" condition when the voltage is more positive than plus three volts with respect to signal ground. The region between plus three volts and minus three volts is defined as the transition region, will be considered in invalid levels. The voltage which is more negative than -15V or more positive than +15V will be considered in invalid levels.

During the transmission of data, the "marking" condition will be used to denote the binary state "one" and "spacing" condition will be used to denote the binary state "zero".

For interface control circuits, the function is "on" when the voltage is more positive than +3V with respect to signal ground and is "off" when the voltage is more negative than -3V with respect to signal ground.

INS8250 Functional Pin Description

The following describes the function of all INS8250 input/output pins. Some of these descriptions reference internal circuits.

Note: In the following descriptions, a low represents a logic 0 (0 volt nominal) and a high represents a logic 1 (+2.4 volts nominal).

Input Signals

Chip Select (SC0, CS1, CS2), Pins 12-14: When CS0 and CS1 are high and CS2 is low, the chip is selected. Chip selection is complete when the decoded chip select signal is latched with an active (low) Address Strobe (ADS) input. This enable comunication between the INS8250 and the CPU.

Data Input Strobe (DISTRDISTR) Pins 22 and 21: When DISTR is high or DISTR is low while the chip is selected, allows the CPU to read status information or data from a selected register of the INS8250.

Note: Only an active DISTR or DISTR input is required to transfer data from the INS8250 during a read operation. Theretore, tie either the DISTR input permanently low or the DISTR input permanently high, if not used.

Data Output Strobe (DOSTR, DOSTR), Pins 19 and 18: When DOSTR is high or DOSTR is low while the chip is selected, allows the CPU to write data or control words into a selected register of the INS8250.

Note: Only an active DOSTR or DOSTR input is required to transfer data to the INS8250 during a write operation. Therefore, tie either the DPSTR input permanently low or the DOSTR input permanently high, if not used.

Address Strobe (ADS), Pin 25: When low, provides latching for the Register Select (A0, A1, A2) and Chip Select (SOC, CS1, CS2) signals.

Note: An active ADS input is required when the Register Select (A0, A1, A2) signals are not stable for the duration of a read or write operation. If not required, the ADS input permanently low.

Register Select (A0, A1, A2), Pins 26-28: These three inputs are used during a read or write operation to select an INS8250 register to read from or write into as indicated in the table below. Note that the state of the Divisor Latch Access Bit (DLAB), which is the most significant bit of the Line Control Register, affects the selection of certain INS8250 registers. The DLAB must be set high by the system software to access the Baud Generator Divisor Latches.

DLAB	A2	A1	A0	Register
0	0	0	0	Receiver Buffer (read), Transmitter Holding Register (write)
0	0	0	1	Interrupt Enable
X	0	1	0	Interrupt Identification (read only)
X	0	1	1	Line Control
X	1	0	0	MODEM Control
X	1	0	1	Line Status
X	1	1	0	MODEM Status
X	1	1	1	None
1	0	0	0	Divisor Latch (least significant byte)
1	0	0	1	Divisor Latch (most significant byte)

Master Reset (MR), Pin 35: When high, clears all the registers (except the Receiver Buffer, Transmitter Holding, and Divisor Latches), and the control logic of the INS8250 Also, the state of various output signals (SOUT, INTRPT, OUT 1, OUT 2, RTS, DTR) are affected by an active MR input. (Refer to Table 1.)

Receiver Clock (RCLK), Pin 9: This input is the 16x baud rate clock for the receiver section of the chip.

Serial Input (SIN), Pin 10: Serial data input from the communications link (peripheral device, MODEM, or data set).

Clear to Send (CTS), Pin 36: The CTS signal is a MODEM control function input whose condition can be tested by the CPU by reading Bit 4 (CTS) of the MODEM Status Register. Bit 0 (DCTS) of the MODEM Status Register indicates whether the CTS input has changed state since the previous reading of the MODEM Status Register.

<u>Note:</u> Whenever the CTS bit of the MODEM Status Register changes state, an interrupt is generated if the MODEM Status Interrupt is enabled.

Data Set Ready (DSR), Pin 37: When low, indicates that the MODEM or data set is ready to establish the communications link and transfer data with the INS8250. The DSR signal is a MODEM-control function input whose condition can be tested by the CPU by reading Bit 5 (DSR) of the MODEM Status Register. Bit 1 (DDSR) of the MODEM Status Register indicates whether the DSR input has changed state since the previous reading of the MODEM Status Register.

Note: Whenever the DSR bit of the MODEM Status Register changes state, an interrupt is generated if the MODEM Status Interrupt is enabled.

Received Line Signal Detect (RLSD), Pin 38: When low, indicates that the data carrier has been detected by the MODEM or data set. The RLSD signal is a MODEM-Control function input whose condition can be tested by the CPU by reading Bit 7 (RLSD) of the MODEM Status Register. Bit 3 (DRLSD) of the MODEM Status Register indicates whether the RLSD input has changed state since the previous reading of the MODEM Status Register.

Note: Whenever the RLSD bit of the MODEM Status Register changes state, an interrupt is generated if the MODEM Status Interrupt is enabled.

Ring Indicator (RI), Pin 39: When low, indicates that a telephone ringing signal has been received by the MODEM or data set. The RI signal is a MODEM-control function input whose condition can be tested by the CPU by reading Bit 6 (RI) of the MODEM Status Register. Bit 2 (TERI) of the MODEM Status Register indicates whether the RI input has changed from a low to a high state since the previous reading of the MODEM Status Register.

Note: Whenever the RI bit of the MODEM Status Register changes from a high to a low state, an interrupt is generated if the MODEM Status Interrupt is enabled.

VCC, Pin 40: +5 volt supply.

VSS, Pin 20: Ground (0-volt) reference.

Output Signals

Data Terminal Ready (DTR), Pin 33: When low, informs the MODEM or data set that the INS8250 is ready to communicate. The DTR output signal can be set to an active low by programming Bit 0 (DTR) of the MODEM Control Register to a high level. The DTR signal is set high upon a Master Reset operation.

Request to Send(RTS), Pin 32: When low, informs the MODEM or data set that the INS8250 is ready to transmit data. the RTS output signal can be set to an active low by programming Bit 1 (RTS) of the MODEM Control Register. The RTS signal is set high upon a Master Reset operation.

Output 1 (OUT 1), Pin 34: User-designated output that can be set to an active low by programming Bit 2 (OUT 1) of the MODEM Control Register to a high level. The OUT 1 signal is set high upon a Master Reset operation.

Output 2 (OUT 2), Pin 31: User-designated output that can be set to an active low by programming Bit 3 (OUT 2) of the MODEM Control Register to a high level. The OUT 2 signal is set high upon a Master Reset operation.

Chip Select Out (CSOUT), Pin 24: When high, indicates that the chip has been selected by active CS0, CS1, and CS2 inputs. No data transfer can be initiated until the CSOUT signal is a logic 1.

Driver Disable (DDIS), Pin 23: Goes low whenever the CPU is reading data from the INS8250. A high-level DDIS output can be used to disable an external transceiver (if used between the CPU and INS8250 on the D7-D0 Data Bus) at all times, except when the CPU is reading data.

Baud Out (BAUDOUT), Pin 15: 16x clock signal for the transmitter section of the INS8250. The clock rate is equal to the main reference oscillator frequency divided by the specified divisor in the Baud Generator Divisor Latches. The BAUDOUT may also be used for the receiver section by typing this output to the RCLK input of the chip.

Interrupt (INTRPT), Pin 30: Goes high whenever any one of the following interrupt types has an active high condition and is enabled via the IER: Receiver Error Flag; Received Data Available; Transmitter Holding Register Empty; and MODEM Status. The INTRPT Signal is reset low upon the appropriate interrupt service or a Master Reset operation.

Serial Output (SOUT), Pin 11: Composite serial data output to the communications link (peripheral, MODEM or data set). The SOUT signal is set to the Marking (Logic 1) state upon a Master Reset operation.

Appendix J

Input/Output Signals

Data (D7-D0) Bus, Pins 1-8: This bus comprises eight TRI-STATE input/output lines. The bus provides bidirectional communications between the INS8250 and the CPU. Data, control words, and status information are transferred via the D7-D0 Data Bus.

External Clock Input/Output (XTAL1, XTAL2, Pins 16 and 17: These two pins connect the main timing reference (crystal or signal clock) to the INS8250.

Programming Considerations

Table 22. Asynchronous Communications Reset Functions

Register/Signal	Reset Control	Reset State
Interrupt Enable Register	Master Reset	All Bits Low (0–3 Forced and 4–7 Permanent)
Interrupt Identification Register	Master Reset	Bit 0 is High, Bits 1 and 2 Low Bits 3–7 are Permanently Low
Line Control Register	Master Reset	All Bits Low
MODEM Control Register	Master Reset	All Bits Low
Line Status Register	Master Reset	Except Bits 5 & 6 are High
MODEM Status Register	Master Reset	Bits 0–3 Low Bits 4–7 – Input Signal
SOUT	Master Reset	High
INTRPT (RCVR Errs)	Read LSR/MR	Low
INTRPT (RCVR Data Ready)	Read RBR/MR	Low
INTRPT (RCVR Data Ready)	Read IIR/Write THR/MR	Low
INTRPT (MODEM Status Changes)	Read MSR/MR	Low
OUT 2	Master Reset	High
RTS	Master Reset	High
DTR	Master Reset	High
OUT 1	Master Reset	High

INS8250 Accessible Registers

The system programmer may access or control any of the INS8250 registers via the CPU. These registers are used to control INS8250 operations and to transmit and receive data.

INS8250 Line Control Register

The system programmer specifies the format of the asynchronous data communications exchange via the Line Control Register. In addition to controlling the format, the programmer may retrieve the contents of the Line Control Register for inspection. This feature simplifies system programming and eliminates the need for separate storage in system memory of the line characteristics. The contents of the Line Control Register are indicated and described below.

Line Control Register (LCR)

3FB

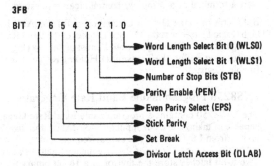

Bit 0 and 1: These two bits specify the number of bits in each transmitted or received serial character. The encoding of bits 0 and 1 is as follows:

Bit 1	Bit 0	Word Length
0	0	5 Bits
0	1	6 Bits
1	0	7 Bits
1	1	8 Bits

Bit 2: This bit specifies the number of Stop bits in each transmitted or received serial character. If bit 2 is a logic 0, 1 Stop bit is generated or checked in the transmit or receive data, respectively. If bit 2 is logic 1 when a 5-bit word length is selected via bits 0 and 1, 1-1/2 Stop bits are generated or checked. If bit 2 is logic 1 when either a 6-, 7-, or 8-bit word length is selected, 2 Stop bits are generated or checked.

Bit 3: This bit is the Parity Enable bit. When bit 3 is a logic 1, a Parity bit is generated (transmit data) or checked (receive data) between the last data word bit and Stop bit of the serial data. (The Parity bit is used to produce an even or odd number of 1's when the data word bits and the Parity bit are summed.)

Bit 4: This bit is the Even Parity Select bit. When bit 3 is a logic 1 and bit 4 is a logic 0, an odd number of logic 1's is transmitted or checked in the data word bits and Parity bit. When bit 3 is a logic 1 and bit 4 is a logic 1, an even number of bits is transmitted or checked.

Bit 5: This bit is the Stick Parity bit. When bit 3 is a logic 1 and bit 5 is a logic 1, the Parity bit is transmitted and then detected by the receiver as a logic 0 if bit 4 is a logic 1 or as a logic 1 if bit 4 is a logic 0.

Bit 6: This bit is the Set Break Control bit. When bit 6 is a logic 1, the serial output (SOUT) is forced to the Spacing (logic 0) state and remains there regardless of other transmitter activity. The set break is disabled by setting bit 6 to a logic 0. This feature enables the CPU to alert a terminal in a computer communications system.

Bit 7: This bit is the Divisor Latch Access Bit (DLAB). It must be set high (logic 1) to access the Divisor Latches of the Baud Rate Generator during a Read or Write operation. It must be set low (logic 0) to access the Receiver Buffer, the Transmitter Holding Register, or the Interrupt Enable Register.

INS8250 Programmable Baud Rate Generator

The INS8250 contains a programmable Baud Rate Generator that is capable of taking the clock input (1.8432 MHz) and dividing it by any divisor from 1 to $(2^{16}-1)$. The output frequency of the Baud Generator is 16x the Baud rate [divisor # = (frequency input) / (baud rate x 16)]. Two 8-bit latches store the divisor in a 16-bit binary format. These Divisor Latches must be loaded during initialization in order to ensure desired operation of the Baud Rate Generator. Upon loading either of the Divisor Latches, a 16-bit Baud counter is immediately loaded. This prevents long counts on initial load.

Divisor Latch   Least Significant Bit (DLL)

3F8   DLAB=1

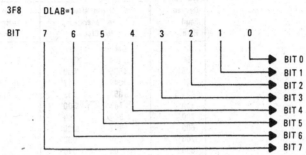

Divisor Latch   Most Significant Bit (DLM)

3F9   DLAB=1

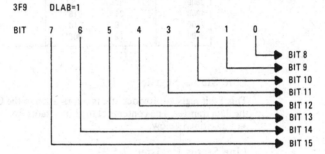

Table 23 illustrates the use of the Baud Rate Generator with a frequency of 1.8432 Mhz. For baud rates of 9600 and below, the error obtained is minimal.

Note: The maximum operating frequency of the Baud Generator is 3.1 Mhz. In no case should the data rate be greater than 9600 Baud.

Table 23.   BAUD RATE AT 1.843 Mhz

Desired Baud Rate	Divisor Used to Generate 16x Clock		Percent Error Difference Between Desired & Actual
	Decimal	Hex	
50	2304	'900'	--
75	1536	'600'	--
110	1047	'417'	0.026
134.5	857	'359'	0.058
150	768	'300'	--
300	384	'180'	--
600	192	'0C0'	--
1200	96	'060'	--
1800	64	'040'	--
2000	58	'03A'	0.69
2400	48	'030'	--
3600	32	'020'	--
4800	24	'018'	--
7200	16	'010'	--
9600	12	'00C'	--

Line Status Register

This 8-bit register provides status information to the CPU concerning the data transfer. The contents of the Line Status Register are indicated and described below.

Line Status Register (LSR)

3FD

BIT   7   6   5   4   3   2   1   0

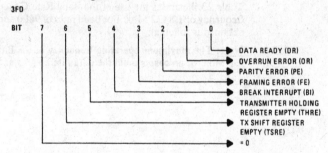

DATA READY (DR)
OVERRUN ERROR (OR)
PARITY ERROR (PE)
FRAMING ERROR (FE)
BREAK INTERRUPT (BI)
TRANSMITTER HOLDING REGISTER EMPTY (THRE)
TX SHIFT REGISTER EMPTY (TSRE)
= 0

Bit 0: This bit is the receiver Data Ready (DR) indicator. Bit 0 is set to a logic 1 whenever a complete incoming character has been received and transferred into the Receiver Buffer Register. Bit 0 may be reset to a logic 0 either by the CPU reading the data in the Receiver Buffer Register or by writing a logic 0 into it from the CPU.

Bit 1: This bit is the Overrun Error (OE) indicator. Bit 1 indicates that data in the Receiver Buffer Register was not read by the CPU before the next character was transferred into the Receiver Buffer Register, thereby destroying the previous character. The OE indicator is reset whenever the CPU reads the contents of the Line Status Register.

Bit 2: This bit is the Parity Error (PE) indicator. Bit 2 indicates that the received data character does not have the correct even or odd parity, as selected by the even parity-select bit. the PE bit is set to a logic 1 upon detection of a parity error and is reset to a logic 0 whenever the CPU reads the contents of the Line Status Register.

Bit 3:This bit is the Framing Error (FE) indicator. Bit 3 indicates that the received character did not have a valid Stop bit. Bit 3 is set to a logic 1 whenever the Stop bit following the last data bit or parity bit is detected as a zero bit (Spacing level).

Bit 4: This bit is the Break Interrupt (BI) indicator. Bit 4 is set to a logic 1 whenever the received data input is held in the Spacing (logic 0) state for longer than a full word transmission time (that is, the total time of Start bit + data bits + Parity + Stop bits).

Note: Bits 1 through 4 are the error conditions that produce a Receiver Line Status interrupt whenever any of the corresponding conditions are detected.

Bit 5: This bit is the Transmitter Holding Register Empty (THRE) indicator. Bit 5 indicates that the INS8250 is ready to accept a new character for transmission. In addition, this bit causes the INS8250 to issue an interrupt to the CPU when the Transmit Holding Register Empty Interrupt enable is set high. The THRE bit is set to a logic 1 when a character is transferred from the Transmitter Holding Register into the Transmitter Shift Register. The bit is reset to logic 0 concurrently with the loading of the Transmitter Holding Register by the CPU.

Bit 6: This bit is the Transmitter Shift Register Empty (TSRE) indicator. Bit 6 is set to a logic 1 whenever the Transmitter Shift Register is idle. It is reset to logic 0 upon a data transfer from the Tranmitter Holding Register to the Transmitter Shift Register. Bit 6 is a read-only bit.

Bit 7: This bit is permanently set to logic 0.

Interrupt Identification Register

The INS8250 has an on-chip interrupt capability that allows for complete flexibility in interfacing to all the popular microprocessors presently available. In order to provide minimum software overhead during data character transfers, the INS8250 prioritizes interrupts into four levels. The four levels of interrupt conditions are as follows: Receiver Line Status (priority 1); Received Data Ready (priority 2); Transmitter Holding Register Empty (priority 3); and MODEM Status (priority 4).

Information indicating that a prioritized interrupt is pending and the type of that interrupt are stored in the Interrupt Identification Register (refer to Table 5). The Interrupt Identification Register (IIR), when addressed during chip-select time, freezes the highest priority interrupt pending and no other interrupts are acknowledged until that particular interrupt is serviced by the CPU. The contents of the IIR are indicated and described below.

Interrupt Identification Register (IIR)

3FA

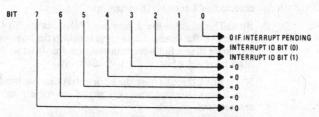

Bit 0: This bit can be used in either a hardwired prioritized or polled environment to indicate whether an interrupt is pending. When bit 0 is a logic 0, an interrupt is pending and the IIR contents may be used as a pointer to the appropriate interrupt service routine. When bit 0 is a logic 1, no interrupt is pending and polling (if used) continued.

Bits 1 and 2: These two bits of the IIR are used to identify the highest priority interrupt pending as indicated in Table 5.

Bits 3 through 7: These five bits of the IIR are always logic 0.

Table 24. Interrupt Control Functions

Interrupt ID Register			Interrupt Set and Reset Functions			
Bit 2	Bit 1	Bit 0	Priority Level	Interrupt Type	Interrupt Source	Interrupt Reset Control
0	0	1	—	None	None	—
1	1	0	Highest	Receiver Line Status	Overrun Error or Parity Error or Framing Error or Break Interrupt	Reading the Line Status Register
1	0	0	Second	Received Data Available	Receiver Data Available	Reading the Receiver Buffer Register
0	1	0	Third	Transmitter Holding Register Empty	Transmitter Holding Register Empty	Reading the IIR Register (if source of interrupt) or Writing into the Transmitter Holding Register
0	0	0	Fourth	MODEM Status	Clear to Send or Data Set Ready or Ring Indicator or Received Line Signal Detect	Reading the MODEM Status Register

Interrupt Enable Register

This 8-bit register enables the four types of interrupt of the INS8250 to separately activate the chip Interrupt (INTRPT) output signal. It is possible to totally disable the interrupt system by resetting bits 0 through 3 of the Interrupt Enable Register. Similarly, by setting the appropriate bits of this register to a logic 1, selected interrupts can be enabled. Disabling the interrupt system inhibits the Interrupt Identification Register and the active (high) INTRPT output from the chip. All other system functions operate in their normal manner, including the setting of the Line Status and MODEM Status Registers. The contents of the Interrupt Enable Register are indicated and described below.

Interrupt Enable Register (IER)

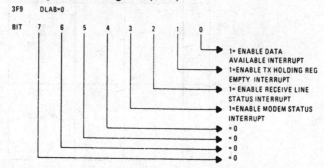

Bit 0: This bit enables the Received Data Available Interrupt when set to logic 1.

Bit 1: This bit enables the Transmitter Holding Register Empty Interrupt when set to logic 1.

Bit 2: This bit enables the Receiver Line Status Interrupt when set to logic 1.

Bit 3: This bit enables the MODEM Status Interrupt when set to logic 1.

Bits 4 through 7: These four bits are alway logic 0.

MODEM Control Register

This 8-bit register controls the interface with the MODEM or data set (or a peripheral device emulating a MODEM). The contents of the MODEM Control Register are indicated and described below.

MODEM Control Register (MCR)

3FC

```
BIT    7   6   5   4   3   2   1   0
                                    └──────────► DATA TERMINAL READY (DTR)
                                └──────────────► REQUEST TO SEND (RTS)
                            └──────────────────► OUT 1
                        └──────────────────────► OUT 2
                    └──────────────────────────► LOOP
                └──────────────────────────────► = 0
            └──────────────────────────────────► = 0
        └──────────────────────────────────────► = 0
```

Bit 0: This bit controls the Data Terminal Ready (DTR) output. When bit 0 is set to a logic 1, the DTR output is forced to a logic 0. When bit 0 is reset to a logic 0, the DTR output is forced to a logic 1.

Note: The DTR output of the INS8250 may be applied to an EIA inverting line driver (such as the DS1488) to obtain the proper polarity input at the succeeding MODEM or data set.

Bit 1: This bit controls the Request to Send (RTS) output. Bit 1 affects the RTS output in a manner identical to that described above for bit 0.

Bit 2: This bit controls the Output 1 (OUT 1) signal, which is an auxiliary user-designated output. Bit 2 affects the OUT 1 output in a manner identical to that described above for bit 0.

Bit 3: This bit controls the Output 2 (OUT 2) signal, which is an auxiliary user-designated output. Bit 3 affects the OUT 2 output in a manner identical to that described above for bit 0.

Bit 4: This bit provides a loopback feature for diagnostic testing of the INS8250. When bit 4 is set to logic 1, the following occur: the transmitter Serial Output (SOUT) is set to the Marking (logic 1) state; the receiver Serial Input (SIN) is disconnected; the output of the Transmitter Shift Register is "looped back" into the Receiver Shift Register input; the four MODEM Control inputs (CTS, DSR, RLSD, and RI) are disconnected; and the four MODEM Control outputs (DTR, RTS, OUT 1, and OUT 2) are internally connected to the four MODEM Control inputs. In the diagnostic mode, data that is transmitted is immediately received. This feature allows the processor to verify the transmit- and receive-data paths of the INS8250.

In the diagnostic mode, the receiver and transmitter interrupts are fully operational. The MODEM Control Interrupts are also operational but the interrupts' sources are now the lower four bits of the MODEM Control Register instead of the four MODEM Control inputs. The interrupts are still controlled by the Interrupt Enable Register.

The INS8250 interrupt system can be tested by writing into the lower four bits of the MODEM Status Register. Setting any of these bits to a logic 1 generates the appropriate interrupt (if enabled). The resetting of these interrupts is the same as in normal INS8250 operation. To return to normal operation, the registers must be reprogrammed for normal operation and then bit 4 of the MODEM Control Register must be reset to logic 0.

Bits 5 through 7: These bits are permanently set to logic 0.

MODEM Status Register

This 8-bit register provides the current state of the control lines from the MODEM (or peripheral device) to the CPU. In addition to this current-state information, four bits of the MODEM Status Register provide change information. These bits are set to a logic 1 whenever a control input from the MODEM changes state. They are reset to logic 0 whenever the CPU reads the MODEM Status Register.

The content of the MODEM Status Register are indicated and described below.

MODEM Status Register (MSR)

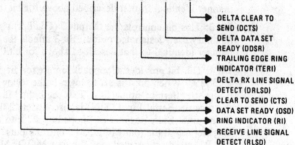

Bit 0: This bit is the Delta Clear to Send (DCTS) indicator. Bit 0 indicates that the CTS input to the chip has changed state since the last time it was read by the CPU.

Bit 1: This bit is the Delta Data Set Ready (DDSR) indicator. Bit 1 indicates that the DSR input to the chip has changed state since the last time it was read by the CPU.

Bit 2: This bit is the Trailing Edge of Ring Indicator (TERI) detector. Bit 2 indicates that the RI input to the chip has changed from an On (logic 1) to an Off (logic 0) condition.

Bit 3: This bit is the Delta Received Line Signal Detector (DRLSD) indicator. Bit 3 indicates that the RLSD input to the chip has changed state.

Note: Whenever bit 0, 1, 2, or 3 is set to a logic 1, a MODEM Status interrupt is generated.

Bit 4: This bit is the complement of the Clear to Send (CTS) input. If bit 4 (loop) of the MCR is set to a 1, this bit is equivalent to RTS in the MCR.

Bit 5: This bit is the complement of the Data Set Ready (DSR) input. If bit 4 of the MCR is set to a 1, this bit is equivalent to DTR in the MCR.

Bit 6: This bit is the complement of the Ring Indicator (RI) input. If bit 4 of the MCR is set to a 1, this bit is equivalent to OUT 1 in the MCR.

Bit 7: This bit is the complement of the Received Line Signal Detect (RLSD) input. If bit 4 of the MCR is set to a 1, this bit is equivalent to OUT 2 of the MCR.

Receiver Buffer Register

The Receiver Buffer Register contains the received character as defined below.

Receiver Buffer Register (RBR)

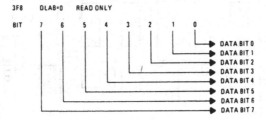

Bit 0 is the least significant bit and is the first bit serially received.

Transmitter Holding Register

The Transmitter Holding Register contains the character to be serially transmitted and is defined below:

Transmitter Holding Register (THR)

3F8   DLAB=0   WRITE ONLY

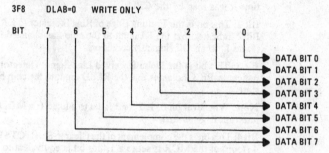

Bit 0 is the least significant bit and is the first bit serially transmitted.

Selecting The Interface Format

The Voltage or Current loop interface is selected by plugging the programmed shunt module, with the locator dot up or down. See the figure below for the two configurations.

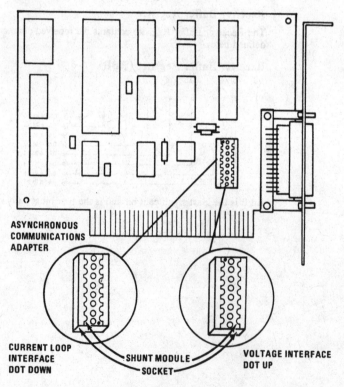

Figure 23. SELECTING THE INTERFACE FORMAT

Asynchronous Communications Adapter Connector
Interface Specifications

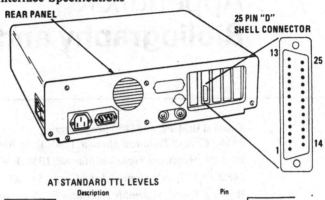

REAR PANEL

25 PIN "D" SHELL CONNECTOR

AT STANDARD TTL LEVELS

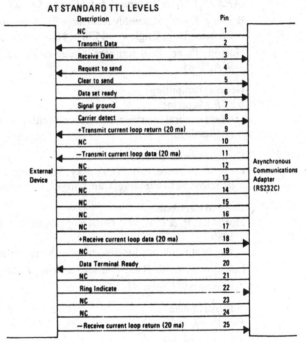

Description	Pin
NC	1
Transmit Data	2
Receive Data	3
Request to send	4
Clear to send	5
Data set ready	6
Signal ground	7
Carrier detect	8
+Transmit current loop return (20 ma)	9
NC	10
−Transmit current loop data (20 ma)	11
NC	12
NC	13
NC	14
NC	15
NC	16
NC	17
+Receive current loop data (20 ma)	18
NC	19
Data Terminal Ready	20
NC	21
Ring Indicate	22
NC	23
NC	24
−Receive current loop return (20 ma)	25

External Device

Asynchronous Communications Adapter (RS232C)

NOTE: To avoid inducing voltage surges on interchange circuits, signals from interchange circuits shall not be used to drive inductive devices, such as relay coils.

Appendix K
Bibliography and sources

General IBM-PC and 8088 programming

IBM-PC DOS Technical Manual, IBM, Boca Raton, FL.

IBM-PC Hardware Technical Manual, IBM, Boca Raton, FL.

iAPX 86,88 User's Manual, Intel Corp., Santa Clara, CA, 1981.

Bradley, David, *Assembly Language Programming for the IBM Personal Computer*, Prentice-Hall, Englewood Cliffs, NJ, 1984.

Norton, Peter, *Programmers' Guide to the IBM-PC*, Microsoft Press, Bellevue, WA, 1985.

General computing

BYTE Magazine.
Good general overview of microcomputing with frequent references to laboratory applications.

General numerical analysis

Press, W., Flannery, B., Teukolsky, S. & Vetterling, W. *Numerical Recipes, The Art of Scientific Computing*, Cambridge Univ. Press, New York, 1986.

General electronics

Horowitz, P. & Hill, W., *The Art of Electronics*, Cambridge Univ. Press, New York 1980.
The best reference for designing laboratory electronics.

Lawson, C. L. & R. J. Hanson, *Solving Least Squares Problems*, Prentice-Hall, Englewood Cliffs, NJ, 1974.

Physical data

Handbook of Chemistry and Physics, ed. R. Weast, 52nd edn, Chemical Rubber Co, Cleveland, OH, 1971.

American Institute of Physics Handbook, ed. D. E. Gray, McGraw-Hill, New York, 1957.

Mark's Standard handbook for Mechanical Engineers, eds. T. Beaumeister, E. A. Abalone & T. Baird, 8th edn, McGraw-Hill, New York, 1978

Physics

Any general introductory physics text will provide a good background.

Sensors and transducers

Doebelin, E. O., *Measurement Systems*, McGraw-Hill, New York, 1983.
A thorough overview of general design and specific devices.

Specific hardware

Witten, I. H., Welcome to the Standards Jungle, *BYTE*, pp. 146–78, February, 1983.
This a close look at serial data communication.

Leibson, S., The Input/Output Primer, Part 3: The Parallel and HPIB (IEEE-488) Interfaces, *BYTE*, pp. 186–208, April, 1982.

Clune, T. R., Interfacing for Data Acquisition, *BYTE*, pp. 269–82, February, 1985.
These two articles provide a good background in how the IEEE-488 works.

General signal analysis

Bendat, J. S. & Piersol, A. G., *Random Data*, Wiley, New York, 1971.

Otnes, R. K. & Enochson, L., *Applied Time Series Analysis*, Wiley, New York, 1978.

Papoulis, A., *Signal Analysis*, McGraw-Hill, New York, 1977.

Specific signal analysis

Monforte, J., The Digital Reproduction of Sound, *Scientific American*, pp. 78–84, December, 1984.
A good description of the sampling problem and digitization.

Cacerci, M. S. & Cacheris, W. P., Fitting Curves to Data, *BYTE*, pp. 340–62, May, 1984.
A description of the Simplex algorithm.

Report writing

Porawn, J. F., *A Student Guide to Engineering Report Writing*, United Western Press, Soloma Beach, 1985.

Hofstaedter, D., Default Assumptions in Metamathecal Themas, *Scientific American*, November, 1983.
For those interested in exorcising the spectre of maleness from their writing.

Index

Index